Tree Breeding and Genetics in New Zealand

C. J. A. Shelbourne • Mike Carson

# Tree Breeding and Genetics in New Zealand

Springer

C. J. A. Shelbourne
Rotorua, New Zealand

Mike Carson
Carson Associates Ltd
Ngongotaha, Bay of Plenty
New Zealand

ISBN 978-3-030-18462-9 ISBN 978-3-030-18460-5 (eBook)
https://doi.org/10.1007/978-3-030-18460-5



This Springer imprint is published by the registered company Springer Nature Switzerland AG
The registered company address is: Gewerbestrasse 11, 6330 Cham, Switzerland

# Preface

As of the year 2018, tree breeding in New Zealand was 67 years old. *Tree Breeding and Genetics in New Zealand* is an account of its development since the late Ib Thulin, the future chief, arrived from Denmark in 1951. Our main motivation for writing this book was the realisation that I (CJAS) was the only tree breeder still surviving who had experienced much of the programme's development (and had contributed much to its thinking and planning). The thinking, the science behind it and the technology have evolved tremendously in that period. This account also demonstrates the international role the New Zealand programme possessed; overseas tree breeders actually came to see how we did it.

'Tree breeding', at the practical level, consists of the selection of superior parent trees to improve their offspring's characteristics that are defined by the breeder. The requirements of selecting 'good' parents, mating them and harvesting their offspring or clonal propagules are common to most domesticated species of plants and animals. The sciences on which this 'tree breeding' is based include genetics, especially quantitative genetics, and statistics. Latterly, genomics has become increasingly important in tree improvement. Plant geography has always been an important basis for planning, especially because all of the species we were improving were exotics, not native to New Zealand.

This book is useful in outlining the early work on provenance differentiation of the many conifers that were tested. A few of these rivalled the species under breeding and selection, but most did not. The tree species that eventually became subjects of breeding programmes were *Pinus radiata, Pseudotsuga menziesii, Pinus contorta, Eucalyptus regnans, E. nitens and E. fastigata*. We will attempt to trace the development of improvement programmes in several exotic tree species through a compendium of abstracts, summaries and summaries of summaries of reports, mostly published. It will also include comments and other contributions by us and the authors.

We see the book being a useful basis for teaching graduate students in forest genetics about the history of tree breeding and tree improvement in NZ. A full-scale provenance testing and plus tree selection programme was initiated in Douglas fir, but it was later realised that the northern coastal populations of the USA were 'wrong' and the breeding programme was put on hold for over 20 years, when the southern coastal USA provenances in the 1959 trials were used to restart the programme. There are many lessons to be learnt from these breeding programmes. Most important is the genetic constitution of the base population, which must include the best-grown and adapted provenances. Another critical choice is what traits to select for, sometimes not done on the basis of genetic parameters. A formal breeding objective is also needed to link trait heritabilities, trait intercorrelations and economic value.

Rotorua, New Zealand C. J. A. Shelbourne
Ngongotaha, Bay of Plenty, New Zealand Mike Carson

# Acknowledgements

Tree breeder Ib Thulin retired in 1983 after a long career dedicated to genetic tree improvement in New Zealand. I joined his team in 1966 and retired in 2006, a work period of 55 years and 66 years till the present. I was aged 32 when I started to work with him at the New Zealand Forest Research Institute (FRI) and left the institute, renamed Scion, when I was 72.

This long period of time means there is opportunity for memory error, but this account will attempt to trace the development of the improvement programme in several exotic species through a compendium of abstracts, summaries and summaries of summaries of reports and publications. It will also include comments and other contributions by me and the authors. Reviews of some aspects of the programme are made in the following chapters.

The NZ Forest Research Institute was more of a research 'village' in the last half of the twentieth century. Everyone knew each other. Ib's office was at the back of what had once been stables (in the 1900s), and my office was up some steps overlooking the old building.

Various staff members all gathered inside or outside the Training Centre cafeteria at the morning and afternoon 'smokos' (tea and coffee breaks). Most lab-oriented groups were in Nissen huts (small portable army huts) to the west of the 'stable' offices. In the 1950s and 1960s, there were no more than half a dozen staff with PhDs in the whole of the FRI.

## Important Colleagues

I don't have all the names of the various technicians who were responsible for establishing the early trials, but colleagues of the time include the following.

## 1950s and 1960s

Eric Appleton, later a South Island nurseryman, and Jap van Dorsser, who was initially involved in radiata pine plus tree selection, were later in charge of the FRI nursery. Trevor Faulds became a vital member of the group, raising seedlings of various species provenances and grafting (for the new seed orchards).

The late Egon Larsen, a friend of Ib from Denmark, was a key person in the early programme. He was particularly good at exploring the local distribution of species, especially in the Pacific Northwest, USA, and organising stand selection and seed collection of desired species. He was no longer with Tree Improvement when I joined, but his name was attached to one of the most important trials, the 1959 Douglas-fir provenance trial.

The late John Miller, a UK-trained forester working in Southland, joined the group in the mid-1950s as a scientist, based in the South Island at Rangiora, where he was later assisted for many years by Robin Parr. Geoff Sweet joined the group in the late 1950s, was producing numerous reports by 1962 and left to do a PhD in tree physiology at Aberystwyth, Wales, in 1964.

I was hired in 1965 while still a student at North Carolina State and started to work with Ib Thulin in late 1966. Mike Wilcox started about the same time as me with a new degree from Oxford, later completing a PhD at NC State in 1973.

## 1970s and 1980s

Mike Carson joined the group in 1976 and completed a PhD at North Carolina State, returning to New Zealand in 1982 with wife, Sue, also with a PhD in forest genetics, who specialised initially in disease resistance selection. The important members of the team included Gerry Vincent, who was originally a forest ranger responsible for trial inspection and maintenance and who later became a specialist in trial establishment and seed orchards, and, later, Tony Firth, who had been a research forester working in tree improvement in Zambia and became the main organiser of the practical side of the breeding programme. Rick Hand was there when I started but later 'migrated' to the Forestry and Timber Bureau in Australia.

There were many others in those early years, including Joe Hignett, Joe Thyne, Irene Steel, Mark Bollmann, Dave Briscoe, Theo Russell, Charlie Low and Jan Riley and later Tony Shorland, John Lee, Neil Woods, Mark Miller, Ruth McConnochie, Toby Stovold, Silvia Concheyro, Dean Witehira and Jodie Wharekura. John Ambrose, a Canadian forester, was involved for the first few years, and Hamish Levack, an NZ forester, was to have an important subsequent career in NZ forestry. Cathy Tosh, a Canadian, later did an advanced degree and eventually was in charge of the black spruce breeding programme in Fredericton, New Brunswick (my UNB alma mater).

## 1990s Onwards

Later, key researchers (several on sabbaticals and postdocs) included Bob Kellison, Randy Johnston, John King, Christine Dean, Luis Gea, Keith Jayawickrama and Satish Kumar. The late Paul Jefferson joined the group from Canada in the early 1990s, prior to his long stint as breeder for the Radiata Pine Breeding Company, and Luis Apiolaza for a brief period, prior to taking up his current professorial role at the University of Canterbury.

Others named in this document are experts who are well-known within their respective fields.

## Images Used in This Book

All images in this book are courtesy of Scion and are used with permission.

## Editorial Assistance

With thanks to Sue Emms for the editorial assistance.

# Contents

# About the Authors

**C. J. A. Shelbourne**

(Chapters 1–15 are authored by Dr. Shelbourne.)

Before coming to New Zealand in 1958, Dr. Tony Shelbourne spent a year researching the ecology of bogs and swamps in New Brunswick, 3 years on species and provenance testing of eucalypts and tropical pines in Zambia and 3 years at North Carolina State University in the USA, studying for a PhD and working on the inheritance of bole straightness and compression wood of loblolly pine (*P. taeda*).

Academic Qualifications

Cyril Joseph Anthony Shelbourne BA Oxon. 1958, MSc New Brunswick 1959, PhD North Carolina State University, 1966

No current memberships of societies

Three Important Papers

C.J.A. Shelbourne 1969: Tree breeding methods. Technical Paper 55, Forest Research Institute, New Zealand Forest Service, Wellington.

C.J.A. Shelbourne, R.D. Burdon, S.D. Carson, A. Firth and T.G. Vincent 1986: Development plan for radiata pine breeding. Forest Research Institute, Rotorua, New Zealand.

C.J.A. Shelbourne, S. Kumar, R.D. Burdon, L.D. Gea and H.S. Dungey 2007: Deterministic simulation of gains for seedling and cloned main and elite breeding populations of *Pinus radiata* and implications for strategy. Silvae Genetica 56: 253–300.

International Consultancy

3806-12711 C.J.A. Shelbourne 1990: A tale of four Coops. A personal overview. (Texas, NC State, Australia and NZ)

4578-13714 C.J.A. Shelbourne 1995: Future opportunities of Carter Holt Harvey in RP breeding: clonal forestry for end product value

6069-15852 C.J.A. Shelbourne 1995: The future development of tree breeding in Chile

C.J.A. Shelbourne 1994 Tree Breeding in sub-tropical China

C.J.A. Shelbourne 1993 South African sabbatical

**Mike Carson**
(Chapter 16 is authored by Dr. Carson.)

Currently director of Carson Associates Ltd., Dr. Michael Carson has worked in the forestry research sector for more than 40 years as, variously, principal scientist, science manager, programme manager, managing director and consultant.

Academic Qualifications

BForSc (Hons – 1st Class), 1976, University of Canterbury

PhD (Forestry, Genetics), 1982, North Carolina State University

Apptd. Honorary Lecturer, 1993, School of Forestry, University of Canterbury

Professional Associations

New Zealand Institute of Foresters, Royal Society of New Zealand

Three Indicative Papers

CARSON, S.D., DJOROVIC N., DJORORVIC A., CARSON M. J., and R. BALL 2003 (submitted Genetics) Simulation of QTL detection and MAS for quantitative traits II: Comparison of gain and selection bias for alternate experimental designs including selective genotyping and map density.

CARSON, S.D.; KIMBERLEY, M.O.; HAYES, J.D.; CARSON, M.J. 1999: The effect of silviculture on genetic gain in growth of *Pinus radiata* at one-third rotation. Can. Jour. For. Res. 29: 1979–1984.

CARSON, S.D.; RICHARDSON, T.E.; CARSON, M.J.; WILCOX, P.L.; DODDS, K.G. 1994: Integrating conventional tree breeding methods with marker aided selection using RAPD markers linked to quantitative trait loci. Proceedings Plant Genome II. San Diego, CA, 24–27 Jan, 1994. Abstract No. P175 (FRI Project Record No. 3937).

International Consultancy

Provision of tree breeding and biotechnology consultancies to forest companies in the USA, Canada, Chile, Argentina, Brazil, PNG, Australia and New Zealand

- Currently designing and implementing a breeding programme for NZ Manuka (Leptospermum sp.) for Comvita NZ Ltd
- Currently designing and implementing a breeding programme for improvement of *Khaya*

- spp. for plantations in Northern Australia and Brazil
- Recent review for Arauco, Chile, validating progress in achieving genetic gain in pine
- plantations
- Reviews of tree breeding programmes in Argentina, PNG, Canada and the USA

# Abbreviations and Acronyms

| | |
|---|---|
| AS | Advanced selection |
| BLUP | Best linear unbiased prediction |
| BRQ | Branch quality |
| BVs | Breeding values |
| DBH | Diameter at breast height |
| CF | Clonal forestry |
| CNI | Central North Island |
| CP | Clonal propagation |
| Cpt | Compartment |
| CRISPR | Clustered regularly interspaced short palindromic repeat |
| GE | Genotype environment |
| GF | Growth and form |
| GxE | Genotype x environment interaction |
| GTI | Genetics and tree improvement |
| GCA | General combining ability |
| GPa | Gigapascal |
| HGT | Height |
| HS | Half-sib |
| IUFRO | International Union of Forest Research Organisations |
| MFA | Microfibril angle |
| MOE | Modulus of elasticity |
| MPa | Megapascal |
| NI | North Island |
| OP | Open pollinated |
| QG | Quantitative genetic |
| Rep | Replications |
| RF | Rep x family |
| RPBC | Radiata Pine Breeding Company |
| RSGCA | Recurrent selection for general combining ability |
| SCA | Specific combining ability |
| SI | South Island |

| | |
|---|---|
| Sib | Sibling |
| SPM | Stems per square metre |
| SSO | Seedling seed orchard |
| Ssp | Subspecies |
| STP | Single tree plot |
| STR | Straightness |
| TI | Tree improvement |

# List of Figures

# List of Tables

# Chapter 1
# The Influence of Ib Thulin

It is a shame that Ib Thulin is no longer in the office next door, where he was, more or less from the time he arrived in NZ until he retired in 1983. He would have explosively corrected the inevitable errors in this account. Ib arrived in NZ in 1951 and in the next one and a half decades accomplished an incredible amount of work, making great progress in the tree improvement of several species. The practical work was done by several field technicians, assisted by John Miller, Geoff Sweet and by Egon Larsen who did much of the species and provenance seed collection in the Pacific Northwest of the USA. Previous experience with using commercial seedlots to characterise provenances had not been good (Fig. 1.1).

Provenance trials in a range of species were created by Ib and his colleagues from 1953 onwards, from designing the sampling, collecting the seed and raising the stock to finding sites and laying out the trials. The designs were apparently all conceived by Ib, and, in some cases, he was able to access seed of provenances of various European species through the generosity of his European tree breeder contacts. He had been trained by Syrach Larsen at Horsholm in Denmark, one of the very few 'fathers' of tree breeding in Europe, and had worked there as a tree breeder for 4 years (with a year's assignment to Greenland) before emigrating to NZ. He had also visited or been in touch with several forest geneticist/tree breeders in Europe and Scandinavia. He was also lucky to have an important backer in New Zealand Forest Service Head Office, Alan Entrican.

The species tested in provenance trials (long-term field experiments) by region included:

Eastern USA: *Pinus (P.) banksiana*, *P. resinosa*, *P. strobus*, *P. taeda*, *P. elliottii*
Europe and Asia: *P. pinaster*, *P. nigra*, *P. sylvestris*, *Picea abies*, *Larix decidua*, *Larix kaempferi (Leptolepis)*
Western USA: *P. radiata*, *P. muricata*, *P. attenuata*, *P. contorta*, *P. ponderosa*, *P. monticola*, *P. lambertiana*, *Pseudotsuga menziesii*, *Tsuga heterophylla*, *Picea sitchensis*, *Thuja plicata*, *Cupressus macrocarpa* (and its hybrid with *Ch. nootkatensis*, Leyland cypress), *Abies grandis*, *Abies* spp.

 
C. J. A. Shelbourne, M. Carson, *Tree Breeding and Genetics in New Zealand*,
https://doi.org/10.1007/978-3-030-18460-5_1

**Fig. 1.1** Cross of clones 850-55 × 850-19 and row of Kaingaroa bulk seedlot p. 1957. Cpt.1038 Kaingaroa

An additional 16 species of Mexican pines were collected by Ib and Egon Larsen in Mexico in 1965 (when he interviewed CJAS in Raleigh, NC for a future job). When later assessed at age 32, the best species were *P. patula*, *P. ayacahuite*, *P. douglasiana*, *P. michoacana*, *P. pseudostrobus* and *P. tenuifolia.*

Martin Bannister's research on progeny and provenance variation in radiata pine started with a 30 family OP test he established in Pigeon Valley near Nelson (results were published in the *NZ Journal of Botany*). A massive series of provenance-progeny trials of radiata pine (from seed collected by Margot Forde, a famous agricultural seed collector) was initiated in 1962 and the first field planting done in 1964. These trials were not included in the TI programme, and Martin remained quite isolated from the TI group.

Rowland Burdon arrived back from Aberystwyth in 1968, and from then on, we became increasingly integrated and eventually became 'Genetics and Tree Improvement', from the previously separate 'Genetics' and 'Tree Improvement'.

The experiments conducted covered native populations, Ano Nuevo, Monterey and Cambria from the Californian mainland, Cedros Island and Guadalupe Island, and two New Zealand (NZ) populations. All five native populations, and two NZ populations, with 50 open pollinated families from each, were planted at two contrasting sites in Kaingaroa. The late Ken Eldridge and Tony Firth carried out a more comprehensive seed collection in the native populations of radiata pine, and these formed a large series of 11 trials across NZ.

Ib started the *Pinus radiata* breeding programme in 1953, focussing entirely on the New Zealand land races that had developed since the first introductions of radiata pine, beginning in the 1860s. Very intensive selection of plus trees was done, mainly in Kaingaroa Forest but with some, including the famous 850-55, selected in Kinleith Forest of NZ Forest Products (based in Tokoroa).

Very large areas (23,000 ha) were searched, but no more than 25 trees were selected, of which only 14 clones were thought good enough for grafting and planting: 10 ha. of seed in the needle-cast orchard in Kaingaroa and another 10 ha. at Gwavas Forest in Hawkes Bay. Both plantings were in 1957.

There was extreme emphasis on intensive phenotypic selection for vigour, an even branching habit, and good stem straightness, but little realisation that narrow sense heritability for many traits might be too low for effective selection. Thus, progeny testing was apparently given little urgency, which subsequently caused problems in the early breeding and seed orchard programme.

In 1960, NZ forestry was still far from being based on a single species. The needle-cast fungus *Dothistroma pini* had not yet arrived, and the Californian provenances of *P. ponderosa* and the Corsican provenance of *P. nigra* were still popular for use on harder sites. *P. contorta* was regarded as an extreme frost-site species for planting at higher elevations. Douglas-fir (*Pseudotsuga menziesii*) was perhaps the most important of the 'alternative' species for softwood timber production. Genetic improvement of these species, however, was still only on the horizon. Eucalypts had not yet been seriously considered as industrial species for solid wood production, though Harry Bunn was pursuing some of these, including stringybark *E. muelleriana.* New Zealand Forest Products Ltd. (NZFPL) would soon establish plantations in their own forests of *E. regnans* and *E. fastigata* for blending with pine fibre for the production of newsprint.

This was the environment in which Ib and his team were operating during the 1950s and early 1960s, an environment that was to change a lot with challenges introduced by the arrival of the needle-cast fungus *Dothistroma pini* about 1965, which more or less eliminated *P. ponderosa* and *P. nigra* as commercial species from the North Island, though they were still usable in low rainfall/higher-altitude areas in the South Island. Dothistroma necessitated a lot of research and practical effort with radiata pine to achieve control.

# Chapter 2
# Species and Provenance Testing of Eastern USA Species

The following summaries are from publications at the NZ Forest Research Institute, Production Forestry Division and Forest Tree Improvement Reports (also called Branch Reports), which are available for inspection in the Scion library. GTI Internal Reports and Project Records are listed in the Appendix and summarised where PDFs of the reports have been made available. The first number in the reference header is the Branch Report number.

**2. I.J. Thulin 1962: Provenance trials of *Pinus banksiana*: establishment report.**
I.J. Thulin's report provided a good review of US and Canadian trials, the earliest from 1940 in Minnesota. There were no reports in the literature of *P. banksiana* trials outside its native area. Records show that 112 acres of this species had been planted in NZ State Forests. Stem form was poor and trees generally unhealthy.

By age 37 years, heights of forest stands in Kaingaroa were 66 ft. and DBH was 7.9 in. Trials were planted in 1955/1956 of five seedlots from Ontario from lat. 45° 30′ to 49° 51′ and one from Quebec at 45° 55′. Seed was sown in 1953 and trials were planted in Kaingaroa and Naseby Forests. Two plots of 225 trees were planted at Kaingaroa and two plots at Naseby of 360–400 trees. Survival at Kaingaroa in 1957 was good (89–92%) and at Naseby in Cpt. 16, but poor in Cpt. 18. Ib's comments were that this was a very limited sample of the species' natural range. 'No extensive trials planned. *P. banksiana* will probably be of no importance as a species for production forestry in NZ, but may have a role in protection forestry'.

**7. G.B. Sweet 1962: A provenance trial of *Pinus banksiana*: first assessment report.**
Survival, height, cone production, foliage colour, crown width, frost damage and animal damage were assessed. Survival was very high, except at Naseby Cpt. 16. Heights averaged 8 ft. at Kaingaroa and 7.7 and 4.6 ft., respectively, at Naseby Cpts. 18 and 16. Provenance 251 from Stevens, Ontario, was consistently the smallest.

 
C. J. A. Shelbourne, M. Carson, *Tree Breeding and Genetics in New Zealand*,
https://doi.org/10.1007/978-3-030-18460-5_2

Generally taller and greener provenances came from milder climates, but growth rates at age 7 of about 1 ft. per year were not acceptable in NZ.

***Comment*** *Pinus banksiana* is essentially a boreal species. With latitudes of 45°, the latitude of Ottawa and Georgian Bay in Eastern Canada, the climate is severely cold in winter, so it was probably a mistake to bother testing this species for use in NZ.

**8. G.B. Sweet 1962: A provenance trial in *Pinus resinosa*: first assessment report.**

Provenance trials of seven seedlots were planted in 1956 at Kaingaroa and Karioi (North Island) and Berwick and Naseby (South Island). Seedlots came from a wide east-west band around the Great Lakes region of Ontario and Michigan, from the east coast to Sault St. Marie in Ontario. Latitudes ranged from 46° 00′ at Petawawa to 49° 25′ at Regina Bay in Ontario. Climates of the various provenances were similar.

There were 2–3 plots per seedlot per site. Survivals were 75% at Kaingaroa, 99% at Karioi, 95% at Berwick and 99% and 86% at Naseby. Mean heights across sites varied from 3.4 ft. (provenance 246 from a more severe climate at Sault St. Marie), and the rest were all 4.1–4.9 ft., with much the slowest growth being at Naseby Cpt. 16. Such slow growth at age 6 did not support use in NZ.

**44. C.J.A. Shelbourne 1970: Growth and morphological properties in a combined provenance and progeny test of slash pine *P. elliottii var elliottii* Engelm.**

In a combined trial, 8 native provenances from the south-eastern USA and 17 open-pollinated progenies from Queensland plus trees of slash pine (originally from Baker Co., FL) were planted in 1956 at Waitangi on a wet, phosphate-deficient gumland clay site. Plot size was 12 × 12 rows or 144 trees, and seedlots were planted in a completely random design with one to three replications (in common with most provenance trials of that time). Also included but not analysed were bulk lots of *P. caribaea var. bahamensis*, *P. caribaea var. hondurensis* and some open-pollinated progenies of *P. elliottii var. densa*.

Systematically located samples of 33 trees from plots with 3 replications, 50 trees from 2 replications and 100 from single plots were assessed for diameter, survival, bole straightness, branching, forking and overall acceptability at age 12 (1968). Some of the provenance seedlots had only single plots. No significant differences among progenies and provenances were detected for DBH. The Queensland progenies were significantly straighter than most provenances, and the best progenies were much straighter than the rest. There were also significant differences among seedlots in branching, survival and forking. The progenies showed significantly higher proportions of acceptable stems than the provenances as a whole. Narrow sense heritability (for the open-pollinated progenies) ranged from 0.39 for straightness, 0.15 for branching to 0 for DBH. Heritabilities are suspect as so few (17) families were involved. Family mean heritabilities were moderate for survival, forking and acceptability. Genetic gains were highest for bole straightness and acceptability.

The other species of *P. caribaea* were greatly inferior to slash pine in diameter growth, and their survival was poor. Top height of the slash pine was about 36 ft. The conclusion was that *P. elliottii var. densa* and both *P. caribaea* ssp. had no commercial future in NZ.

Growth of slash pine was promising, considering the poorly drained and P-deficient site at Waitangi. The best Queensland progenies (from northern Florida, Baker Co.) were considerably superior to the rest, and clonal orchard seed of these would yield twice the gain of the OP progenies. A seedling seed orchard of orchard parents would be a desirable basis for any slash improvement programme for Northland.

***Comment*** These remarks were made in 1970 about a 12-year-old trial and are now (2018) completely outdated, on the basis of wood properties, subsequent growth and the superior performance of *P. radiata* with good land preparation and fertiliser. I don't think anyone would consider planting slash pine in 2018.

**54. C.J.A. Shelbourne 1971: Provenance variation in growth rate and other characters in 13-year-old loblolly pine (*P. taeda L.*) in New Zealand.**
This trial was sown in 1954, planted in 1955 at Rotoehu and Waitangi Forests and assessed at age 13 years from planting. It was one of the earliest provenance trials established by Ib. The seed came from Jim McWilliam in Queensland and duplicated the trial planted there. The layout was a completely random design, with one to three plots of 196 trees of each seedlot and was planted at Waitangi and Rotoehu. The seedlots were divided into seven groups, according to origin and geography, namely:

1. Louisiana provenance
2. Five select open-pollinated progenies and one bulk from Queensland
3. Three Georgia and South Carolina provenance lots
4. Two Arkansas provenance lots
5. Rotoehu, NZ, three progenies bulked and South Africa bulk
6. Two North Carolina and Virginia provenance lots
7. Maryland provenance

Samples of 33, 50 or 100 trees per seedlot were assessed at each site for the following traits depending on the number of plots per seedlot: diameter at breast height (DBH), forks (presence or absence), straightness (1–10), branching (1–10), acceptability for pruning, survival and wood density (20 per seedlot). Top height was estimated from the tallest ten trees per plot. Best basal area growth was from Group B (Queensland) followed by C, D, E (NZ), F and Maryland last.

NZ lots from Rotoehu resembled the straight-stemmed Arkansas provenance. The Louisiana provenance was poorly stocked and consequently ranked best for DBH but lower for top height. Wood density was very low and was best for the Arkansas and Rotoehu lots which also showed much the highest acceptability scores. Density averaged 350 kg/m$^3$ at Rotoehu. The bark pocket problem (pockets of bark in characteristic infoldings above and below branches) was only mentioned in the Introduction (see Harris and Birt Silvae Genetica in press).

***Comment*** In retrospect, the report was rather upbeat, reflecting the fact that at the time *P. taeda* was still considered a possible species for the P-deficient sites in Northland. The later malformation caused by possums at Rotoehu really decimated the species in spite of reasonable volume growth (no possums at Waitangi then). The bark inclusions and low wood density plus lack of summerwood development were not emphasised sufficiently and later contributed to the species demise as a commercial prospect.

**PR 510 PROPPS 0300 C.J.A. Shelbourne 1984: Choosing the best provenance of Loblolly pine (*P. taeda L.*) for New Zealand.**
A final assessment (at age 28 years) of the trial described above showed similar differences among provenances in growth rate and bole straightness. The Arkansas provenance was still by far the best for straightness, but the coastal provenances from South Carolina, Georgia and Florida (via Queensland) were superior for growth rate. Some Queensland progenies had straightness ratings almost as high as Arkansas but with superior volume growth.

In 1971 (13 years previously), I came to some rather sweeping conclusions about the future of the species and how to handle it as a contingency species. It was proposed to establish seed stands of seed from seed orchards in South Carolina, Georgia, Florida and Queensland.

Although quite extensive plantings were made of Southern pines from 1920 to 1960 at Rotoehu and various Northland forests where P-deficient gumland clay soils predominated, the spread of opossums northwards ruined the form of young *P. taeda*. Radiata also performed well with phosphate fertilising on these sites. The *P. taeda* wood properties were inferior to radiata's, with summerwood lacking and bark pockets prevalent beneath branches. In retrospect, *P. taeda* looks a very poor prospect as a contingency species.

**IR PROPPS C.J.A. Shelbourne and I.J. Thulin 1976: Provenance variation in *Pinus strobus*: heights five years after planting in New Zealand.**
Seventy-seven provenance seedlots of *P. strobus* were out-planted in 1970 as 1/3 stock at three sites, Gwavas, Rotoehu and Golden Downs. Five seedlots of *P. monticola* and one of *P. griffithii* were also included. This was one of the first applications of a 'sets in replications' design in NZ for provenance trials, with 20 provenances per set and 5 reps of 10-tree plots.

Heights 5 years after planting showed that southern Appalachian provenances (from North Carolina, South Carolina, Georgia and Kentucky) grew the fastest, followed by provenances from the Central Eastern group (West Virginia, Virginia, Pennsylvania and Maryland) intermediate and New England provenances, and those from Manitoba, Minnesota, Wisconsin, Ontario, Quebec, New Brunswick and Nova Scotia, growing much slower. The generally slow growth, relative to *P. radiata*, frequent multiple leaders and very low wood density make this species unpromising in New Zealand.

**IR PROPPS Chen Jianzin 1989 Provenance selection of *Pinus strobus* in New Zealand.**

Some 3000 ha. of *P. strobus* were planted, mostly in the 1920–1930s with seed from Ontario (about the worst provenance, as this trial indicates). About 85% of this planting was in the NI. Little planting has been done since. A nationwide provenance trial with 77 seedlots covering much of the natural range of this species was planted at Gwavas, Rotoehu and Golden Downs in 1970, but Golden Downs was not assessed at age 18 years. Traits assessed were diameter at breast height (DBH) and forking percentage.

The design was a modification of sets in reps, with seedlots divided into sets of about 20 each and with a mixture of provenances/progenies that was similar in each set. Five replications of the 10-tree plots of each set were planted per site, but sets C and D had only sufficient stock to be planted at Gwavas. This planting strategy was a radically different design than had been used for provenance testing in the 1950s and 1960s. We went to the other extreme with reduced numbers of trees per plot, partly because with so many provenances (77), a general picture of regional genetic variation was needed.

Highly significant differences were found for DBH and forking percentage in analysis of every set. Provenance × replication variance was negligible. Provenance × location variance was significant between Gwavas and Rotoehu for DBH and forking, although F ratios were relatively small.

There was a strong trend of decrease in DBH growth from the southern Appalachians northwards; states ranked Ga., NC., SC., Tenn., Kent., WVa., Va., Md., Penn., Mich., NY., NS, Ont., Que., Minn. and Manitoba.

## References

2. I.J.Thulin 1962: Provenance trials of *Pinus banksiana*: establishment report
7. G.B. Sweet 1962: A provenance trial of *Pinus banksiana:* first assessment report
8. G.B. Sweet 1962: A provenance trial in *Pinus resinosa*: first assessment report
44. C.J.A. Shelbourne 1970: Growth and morphological properties in a combined provenance and progeny test of slash pine *P. elliottii var elliottii* Engelm.
54. C.J.A. Shelbourne 1971: Provenance variation in growth rate and other characters in 13-year-old loblolly pine (*P. taeda L.*) in New Zealand
PR 510 PROPPS 0300 C.J.A. Shelbourne 1984: Choosing the best provenance of Loblolly pine (*P. taeda L.*) for New Zealand.
IR PROPPS Chen Jianzin 1989 Provenance selection of *Pinus strobus* in New Zealand.

# Chapter 3
# Species and Provenance Testing of European and Japanese Species

**3. G.B. Sweet 1962 Provenance trials of *Pinus pinaster*: first assessment report.**
Seedlots of this widely distributed species were sown in 1953 and trials planted in 1955 on nine sites. Like *P. banksiana* and *P. resinosa*, this species was part of Ib's first set of trials, sown a couple of years after his arrival in NZ. However, the plantings represented a much better 'punt' on a useful species for NZ forestry. The report covers assessment in 1960 of a major trial and should have been published after some restructuring and abbreviation. At the time *P. pinaster* (Maritime pine) was an important alternative species.

There was a good literature review: 47 seedlots were included, sown in 1953 (an additional 4 lots were sown in 1954). Stock was raised as 1 + 1. Nineteen seedlots were from Portugal ('Atlantic'), four from Spain, four from France ('Atlantic'), eight from Italy, five from Corsica, three from Morocco and two from NZ. Stock for South Island trials were lined out at Milton, near Dunedin.

The experimental design at Woodhill and Waitangi was randomised blocks, probably with a completely random layout, with two or three replications of 196-tree plots per site. At Rotoehu, Berwick, Puhipuhi, Riverhead and Naseby, single 100-tree plots of each seedlot were planted, unreplicated.

Traits assessed were height, bole straightness (acceptable or nonacceptable), leader dominance (acceptable or nonacceptable), needle length, coning, survival, crown width and damage. At Woodhill, 36% (about 70 trees per plot) were measured and assessed for height, straightness, dominance and survival, 50% (about 50 trees per seedlot) were assessed at Puhipuhi and Rotoehu and 100% at Naseby and Berwick, for height and survival only. Results are tabulated by site and seedlot (accessed under Branch Report 3 in Scion library) at Woodhill, Rotoehu and Puhipuhi and are comprehensive.

On the four sites assessed in the North Island, Atlantic coast origins showed much faster height growth than other origins from Italy, Corsica and Morocco but were of lower survival and poorer stem straightness. In the South Island, at Naseby



C. J. A. Shelbourne, M. Carson, *Tree Breeding and Genetics in New Zealand*,
https://doi.org/10.1007/978-3-030-18460-5_3

and Berwick, height growth was slow and Atlantic origins did not appear well suited. Corsican and inland Spanish origins were the best.

In spite of the very large number of trees per plot, having few replications meant that seedlot differences were often undetectable (50–70 trees per plot were assessed). Within-plot site variation and other plot-to-plot effects predominated. However, when seedlots could be grouped, e.g. France Atlantic versus Portugal Atlantic, precision was improved – provided there was little genetic variation within regions and there were at least five seedlots per region. In the long term, the big plots gave longevity to the trials.

Unfortunately, this assessment at age 5 years was never published. A later assessment at Woodhill (in which I helped), at about age 20, by Tony Firth, was probably never written up.

Differences in height among seedlots at age 5 gave rise to the following groupings:

(a) Atlantic: Portugal, NW coast of Spain and the Landes of France (tallest but poor straightness)
(b) Genova-La Spezia, Italy
(c) Lucca-Esterel, France and Italy, quite straight
(d) Corsica, very straight stems, best leader dominance. Best growth at Naseby
(e) Morocco, quite straight
(f) Inland Spain, good at Berwick

Much tree improvement work was conducted with this species in Western Australia, starting in the 1940s, and a tree improvement programme existed there from 1950 onwards. Perry spent many years in Portugal selecting plus trees and shipping scion material back to WA. He also sent scion material to NZ, and two clonal seed orchards were planted, including a small one in the Long Mile research area, in Rotorua. *Pinus pinaster* was an important species in the Cape Province of South Africa till about 1990, when its place was taken over by *P. radiata*.

***Important contacts*** are Hopkins, Perry, Rycroft and Wicht, Scott and Colin Duff. It was interesting that I worked in Zambia from 1960 to 1963 on tropical pine provenance testing, and Colin Duff, in 1940, had planted a block of *Pinus insularis* (*P. kesiya*) near Ndola on which afforestation with this species was later based.

**I.J. Thulin. ca. 1954 *Pinus sylvestris* provenance research.**
The trials were planted in 1956. There were five provenance seedlots only, all from Spain, and they were provided by Fernando Molina (Ib's contact). They came from 3600 to 4900 ft. altitude and varying latitudes from 37° (Auckland/Cadiz) to 42° (south of Nelson/Barcelona) with wide distribution from the Pyrenees to the far south in Grenada. Spanish provenances were not represented in other European trials, and Ib must have made the decision not to explore the species further. The experimental design was minimal, and trials were planted in 1956 at Kaingaroa Cpt. 634 (altitude 2300 ft.), Berwick Cpt. 36 (altitude 100 ft.) and Naseby Cpt. 18

(altitude 2200 ft.). Only one plot per provenance was planted at each site, 100 trees at Kaingaroa, 121 at Berwick and 190 at Naseby. Survivals at 1958 age 2 years were all near 100%. This species was 'not considered to have any great potential for production forestry in NZ'. Later it did rather well in high-altitude species trials. But now exotic species for protection are no longer seen as appropriate.

The five seedlots came from a wide distribution in Spain, as shown above. Survivals were excellent up to 1962, age 6 years, and were all in the 1990s except at Kaingaroa where the lowest was 80–85%. Heights at age 6 varied from 3.3 to 4.7 ft. at Kaingaroa, 4.3–6.7 ft. at Naseby and 5.7–7.7 ft. at Berwick. Seedlot 261 from the Sierra Nevada in Grenada (far south) had slowest growth at all sites. There is no means of relating these rates with those of other provenances, but reported that a Golden Downs stand of Spanish provenance was 46 ft. tall at age 23. There was later interest in *P. sylvestris* for high-altitude planting in the South Island.

**5. I.J. Thulin 1962 Provenance trials of *Larix decidua and L. leptolepis*: Establishment Report.**

Two series of provenance trials were planted: in 1957, *L. leptolepis* (Japanese), and in 1959, *L. decidua.*

*Larix leptolepis Trials*

There were 20 *L. leptolepis* seedlots sown in 1955 at the Forestry Research Institute, which were later lined out at Golden Downs and Milton. The 15 seedlots of *L. leptolepis* (one NZ lot) were all from the island of Honshu, with altitudes varying from 1320 to 2000 m, MAT varied from 3.3 to 6.5 °C and rainfall from 1360 to 2840 mm. Seedlots were provided originally by Dr. Iwakawa to Drs Langner and Schober in Germany.

There were 1–3 plots with 99 trees/plot at Kaingaroa Cpt. 634, 1–3 plots of 196 trees at Kaingaroa Cpt. 1149 and 1–3 plots of each seedlot with 121 trees/plot and single plots at Patanamu, Gwavas, Golden Downs, Hanmer, Berwick and Naseby.

*Larix decidua Trials*

There were 27 native population seedlots of *L. decidua,* from France, Italy, Austria, Czechoslovakia and Poland, and 10 exotic seedlots included 3 from NZ in the 1959 series. The *Larix decidua* trials were planted at Kaingaroa Cpt. 634 (1–2 plots per seedlot, 100 trees/plot) and Kaingaroa Cpt. 1149 (1–2 plots per provenance, 140 trees/plot). Other trials at Patanamu, Gwavas, Golden Downs, Hanmer, Berwick, Rankleburn, Naseby and Whaka had 1–3 plots per site and 121–144 trees per plot.

Survivals 1 year after planting were nearly all very high, in the 90%. The shorter provenances came from the western Alps, taller ones from the Sudeten Mountains in Czechoslovakia, NZ and German Schlitz-Hessen, while Tatra and Polish provenances were intermediate. Heights were greater at four South Island (SI) sites than on three North Island (NI) sites. Trees were straighter from Scottish and French Alpine seedlots, with poorer straightness of NZ, Sudeten and German provenances.

This early assessment report had a good literature review, especially of the early European literature.

**17. Miller, J.T. and Fairburn, H.S. 1963. Provenance trials in *Larix decidua and Larix leptolepis*. First assessment.**
The 1957 series of 20 seedlots of *L. leptolepis* were sown in 1955 at FRI and later lined out at Golden Downs and Milton.

*Larix decidua*

There were 37 seedlots of *L. decidua*, 27 native population seedlots from France, Italy, Austria, Czechoslovakia and Poland and 10 exotic seedlots, including 3 from NZ. The provenances from the western Alps were shorter and were taller from the Sudeten Mts in Czechoslovakia, NZ and German Schlitz-Hessen. Tatra and Polish provenances were intermediate. Heights were taller at four SI (South Island) sites than on three NI (North Island) sites. Straighter trees came from Scottish and French Alpine seedlots; trees with poorer straightness were from NZ, Sudeten and German lots. Provenance mean heights ranged from 11.6 ft. for best NZ lots to 5.2 ft. for Austrian and French alpine lots.

Growth was best at Golden Downs and Berwick and slowest at Gwavas and Hanmer.

There was a good literature review of (ancient) European literature.

**11. J.T. Miller 1962. The problems of forest tree improvement in larch in New Zealand. Problem analysis.**
This problem analysis should have been made much earlier, before the major efforts of provenance research and, later, selection and breeding were started. Is genetic improvement of larch (Japanese or European) really warranted and to what extent?

There were about 9300 acres of *L. decidua* and 700 acres of *L. leptolepis* in NZ in 1962. First introduced in 1886, *L. decidua* was regarded as important until about 1915. It was acknowledged to be the strongest exotic softwood, but it had been inadequately tested for provenance. Larch came to be regarded as just an amenity species. The interspecific hybrid European × Japanese had been popular overseas, and there was a long literature review of hybrid larch (35 references) in this problem analysis. 'A little work done in the right place' (Miller 1962) was proposed for NZ, via hybrids of selected provenances and select parents.

***Comment*** This document proposes work with little basis for the use of the hybrid product and is poorly conceived in spite of the thoughtful introduction. With current perspectives, this breeding programme should never have been started.

**28. J.T. Miller 1965. Provenance trial in *Larix leptolepis*: assessment 1964.**
There were 15 seedlots (one NZ lot) of Japanese larch, all from the Island of Honshu (see Branch Reports 5 and 17). Other trials were planted by J.W. Wright in the USA.

One-to-three plots of each seedlot were planted at nine sites, with 99–296 trees.

At the assessment, 5 years after planting, survivals were generally high, 65–93%. Tallest provenances were from Azusayama, Kurokchi and Yatsutake forests and Mt. Fuji (7.9–11.4 ft.). There were no strong correlations between height and altitude, MAT or mean precipitation.

**29. J.T. Miller 1965 Provenance trial in *Larix decidua*. Second assessment report. Age 6 years.**

At age 5 years, 27 native population seedlots of *Larix decidua* from France, Italy, Austria, Czechoslovakia and Poland were measured, plus 10 exotic seedlots (including 3 from NZ) (Miller & Fairburn Report 17 of 1963).

This was a very large set of trials with good geographic coverage of native populations at nine or more NZ sites. Ranking of provenances is almost impossible at any individual site (with number of plots per seedlot varying from 1 to 3). With good early survivals and large plots, the trials should be long lasting (Fig. 3.1).

The results are from 5 years after planting of 8 sites of 11 provenances of European and 2 of hybrid larch. Survivals were high, 72–96%, and there were small differences between provenances. Provenance height means ranged from 6 to 11 ft., and the hybrid seedlot was the tallest at 13 ft. Polish seedlots as a group were taller than the rest. A sample of 50 trees per plot, 13 seedlots (including the hybrid) and 8 sites was assessed. There was significant provenance × site interaction which may have derived from the uneven numbers of plots for each provenance at each site. This was a least squares analysis with estimates of individual provenance and site effects.

**Fig. 3.1** European larch selection

There were no estimates of provenance means at each site as there was little replication. Site means ranged from 6.9 ft. at Golden Downs to 12.9 ft. at Hanmer. These results can be compared with those in report 17: 'shorter provenances were from the western Alps, taller from Sudeten Mountains in Czechoslovakia, NZ and German Schlitz-Hessen, with Tatra and Polish provenances intermediate'. This pattern of best height growth by Polish-Czechoslovakian provenances was reported from Scotland and from the USA. Hybrid European × Japanese larch always grew fastest, though how much of this superiority was due to high GCA of the parents is not known (see GTI (Genetics and Tree Improvement) Reports 5 and 17, also Edwards, Scottish studies of the provenances of European larch IUFRO (International Union of Forest Research Organisations) Cong. Proc. 11 (1953) 432–437.

**64. J.T. Miller 1973 An extraction thinning in Larch provenance trials.**
Two adjacent trials were planted at Hanmer: 15 seedlots of *L. leptolepis*, planted in 1957, and 17 seedlots of *L. decidua* and one of the hybrid *L. decidua* × *L. leptolepis*, planted in 1959. This thinning in 1973, age 14 years, reduced stocking of plots from 6 × 6 ft. (1.8 m, 3090 stems/ha) to 1000 stems/ha. or 40 per plot.

Extractions from hybrid plots were 82.9 $m^3$/ha., versus 29.5 $m^3$/ha. from *L. decidua* and 27.9 $m^3$/ha. from *L. leptolepis* plots. Residual volumes per hectare were, respectively, 94.3, 67.5 and 65.8 $m^3$. The hybrid plots were straighter and yielded a larger proportion of 'stays'. All plots were reduced to 40 trees. Superiority of hybrids was due to their greater size and the selection of the probably outstanding female *L. decidua* parent in Denmark. The male Japanese parents were also two that were superior in progeny tests in Denmark though Japanese parents showed no difference from the European in volume growth at Hanmer. Best growth in NZ was from exotic provenances from Denmark and Germany and from Polish mountain provenances.

**16. J.T. Miller 1964 Provenance trial of *Pinus nigra*: establishment report.**
The seedlot origins were Spain, seven; France, three (Cevennes); Italy, four (two Calabrican); Corsica, four; Austria, two; Yugoslavia, ten; Turkey, five; Greece, one; and Cyprus, one. Among the plantation-derived seedlots, there were NZ, four; England, three; France, one; and Denmark, one (one *P. halepensis* from Turkey).

This was a major provenance trial of a then-important species. Seed weights were highly variable, from 14 to 27 g, high for Yugoslavia, Turkey and Austria, lowest for Cyprus. Nursery heights ranged from 10.7 to 3.2 in. Survivals in the nursery were generally high (76–94%) but lowest for Corsican seedlots and highest for Spanish and Italian. NZ lots were similar in appearance to Corsican, taller than average and with best survival.

In 1856–1958, 47 *P. nigra* seedlots were planted at 9 sites: 5 sites in the NI and 4 in the SI. Plantings were assessed for height and survival at eight sites and crown width at one. At most trials, one plot per seedlot was planted, with three plots planted only at Karioi and at Golden Downs. One-hundred trees/plot were planted at Kaingaroa, Gwavas, Ngaumu, Hanmer, Naseby and Berwick. One-hundred forty-four trees per plot were planted at Karioi and Golden Downs. Twenty-five trees/plots were planted at Whaka.

**30. J.T. Miller 1965 Provenance trial in *Pinus nigra*. First assessment report.** In NZ trials the seedlot origins were seven from Spain, three from France (Cevennes), four from Italy (two Calabrican), four from Corsica, two from Austria, ten from Yugoslavia, five from Turkey, one from Greece and one from Cyprus. Among the plantation-derived seedlots, there were four from NZ, three from England, one from France and one from Denmark (one *P. halepensis* from Turkey). Planting was done over 3 years, 1956, 1957 and 1958, at nine sites. Survival was generally good at 76–94% over sites, best in Spanish and Italian seedlots and worst in Corsican, Greek and Cyprus provenances.

Trials were planted in 1956–1958 of 47 *P. nigra* seedlots which were assessed for height and survival at eight sites and crown widths at one site at age 8–9 years. *P. halepensis* was measured at three sites. Tallest seedlots were from the west of the natural range, that is, Spain, France, Sicily and Corsica. Shortest seedlots were from the east, Yugoslavia, Turkey, Greece and Cyprus. Nine explantation seedlots all showed above-average growth, and four NZ lots were equally tall with the Spanish. The shortest branches were found in the Corsican and most plantation-derived seedlots, including NZ lots.

There was good coverage of taxonomy and provenance research overseas. British experience was that Corsican origins grew faster than Austrian and Spanish, though Austrian seedlots were better on limestone and chalk. Differences among provenance-region mean heights at ages 8–9 years were meaningful though seedlot (provenance) means at each site were generally meaningless with so little plot replication. Provenance-region mean heights across sites were as follows: Spain 5.5 ft., France 4.9 ft., Sicily 4.0 ft., Italy 3.9 ft., Corsica 4.6 ft., Austria 3.7 ft., Yugoslavia 3.7 ft., Turkey 3.7 ft., Cyprus 2.9 ft., Greece 3.5 ft., NZ 5.3 ft., the UK 4.4 ft., France 4.5 ft. and Denmark 5.2 ft. Crown width/height ratios were Corsica 0.50, NZ 0.51, England 0.50, Denmark 0.48 and the rest 0.65–0.71.

This was a remarkably clear result, indicating a major separation of the Corsican race from the rest and the source of UK, NZ and Danish seedlots. At the beginning of State plantations, an early distinction was made between *P. nigra* sub-species *austriaca* and Corsican sub-species *laricio*. This sub-species ranked third in annual planting area behind larch in the early 1900s. Planting of ssp. *austriaca* generally ceased after 1925.

**68. M.D. Wilcox and J.T. Miller 1974 *Pinus nigra* provenance variation and selection in New Zealand (see same title Silvae Genetica 24 (5–6): 132–140).** *Pinus nigra* was one of the first exotic timber species used in plantations in NZ, 20–30 years before seed of *P. nigra* was collected from apparent Corsican pine at Dumgree in Marlborough and a further 2610 kg imported from Corsica. In 1970 (some 5 years after Dothistroma was first detected), *P. nigra* stands were 30,000 ha., the third ranked exotic species. Little new planting has been done since, because of Dothistroma, apart from some in Dothistroma-free areas in the interior of the SI where it still looks good.

There is a good review in this paper of the sub-species and their natural distribution in Europe.

NZ provenance trials include 34 seedlots from native stands, 5 from (exotic) European plantations and 4 from NZ stands. Sowing was in 1954 and 1955 and the trials were planted in 1956 and 1958. The main trials were planted at four sites in the NI and four in the SI. Karioi and Ngaumu were assessed in the NI and Golden Downs, Hanmer, Naseby and Berwick in the SI.

Thirty randomly selected trees per seedlot per site were assessed at age 16 years for height, DBH, stem straightness, forking and stem volume (30 trees for single plots, 2 × 15 for 2 plots and 3 × 10 trees for 3 plots). This sampling reflects previous experience that showed assessing large numbers of trees from a few plots was a waste of effort.

Analysis of provenance means was done at six sites, with missing provenance means calculated by least square estimates. Analyses of variance mean squares were provenances, locations, provenances × locations, plots in locations (cells) and trees in plots.

Provenance × location interaction was significant in all traits, but there were no rank changes of sub-species groups. The significant interaction possibly reflects the variable number of plots per provenance per site. There was one plot per seedlot per site of all provenances, except at Karioi and Golden Downs where there were mostly three. Sub-species *laricio* from Corsica and NZ seedlots were top-ranked for height at all sites though superiority was least at Naseby. Other native ssp. *laricio* lots include Sicily and Calabria and exotic, from the UK and Denmark which ranked second for height. Sub-species *salzmannii* (Spain) ranked lowest (though fastest at age 5) along with ssp. *nigra* (Italy, Yugoslavia and Austria) and other ssp. *palassiana* which ranked above Spain. Bole straightness (conveniently) showed the same rankings.

There was no possum damage on all ssp. *laricio* (originally Corsican plus Calabrica and Sicily), but all other provenances were severely damaged by possums. Branch habit of Corsican was much lighter, and the stem forms much better than other sub-species. Native and exotic Corsican seedlots showed little frost damage, while other provenances did.

However, the Corsican provenances are highly susceptible to *Dothistroma pini*, and this combined with their slower growth than radiata pine prohibits their use except perhaps in Dothistroma-free and harsh climate areas in the interior of the SI. There were a few well-grown but poor-straightness provenances from Spain.

This is an interesting and instructive history of this species, with its wide European distribution and large genetic variation in many different traits. It is sad that a disease, Dothistroma, has completely put paid to it in favour of radiata, which of course grows much faster. In a ‘radiata dead’ scenario, it would probably be possible to create a Dothistroma-resistant inter-provenance hybrid with the better *P. nigra* provenances. Its form, growth and hardiness still impress in harsh inland climates in the SI where Dothistroma is not a problem. It would then have to compete with a *P. attenuata* × *P. radiata* hybrid.

A selection index with volume, straightness and forking was developed which ranked the Corsican lots, native and exotic in the upper part, with the Dumgree lot top and ssp. *nigra* and *pallasiana* (Austria and Cyprus) bottom, with ssp. *salzmannii* somewhat intermediate (France, Italy, Turkey).

**D.J. Cown 1974: Physical properties of Corsican pine grown in NZ. NZJFS (4).** The physical properties of Corsican pine (*Pinus nigra*, Arnold, most likely ssp. *laricio*) were examined using increment cores and wood discs from 41 sites. Two main classes of stand were sampled: grown from seed imported from overseas in the 1920s (lots HO 26/1, HO 27/27 and HO 28/112) and second-generation stands from seed collected in Dumgree, New Zealand (lots NM 46/427 and NM 47/451). Apart from the age differences between the two groups, variation between seedlots was found to be small, probably due to the fact that crops grown in New Zealand seem to be of mixed var. *calabrica* and var. *austriaca* origins (*comment: this is wrong. NZ stands are ssp. laricio with some stands of ssp. austriaca*).

Wood density was consistently higher than that found in other commercially grown exotic conifers, and outerwood densities were observed to decrease with increasing altitude and latitude. A strong correlation was found between outerwood density at breast height and tree mean density. Resin content was very high, particularly in the heartwood, where it was over 20% in individual stems. This will affect the economics of pulping the older crops. However, heartwood development commenced late and progressed slowly. Tracheid lengths were intermediate between those previously found for radiata pine and lodgepole pine.

***Comment*** This was generally a good resource-based study of a wide variety of NZ stands at the time, (though I believe he got the taxonomy wrong). With outer 10 rings on 50 trees per site, this was a big study with 41 sites but rather few seedlots.

## References

3. G.B. Sweet 1962a Provenance trials of *Pinus pinaster*: first assessment report.
6. G.B. Sweet 1962b. Provenance trial in *Pinus sylvestris*: First Assessment report.
5. I.J. Thulin 1962 Provenance trials of *Larix decidua and L. leptolepis*: Establishment Report.
17. Miller, J.T. and Fairburn, H.S. 1963. Provenance trials in *Larix decidua* and *Larix leptolepis*. First assessment
11. J.T. Miller 1962. The problems of forest tree improvement in larch in New Zealand. Problem analysis.
28. J.T. Miller 1965a. Provenance trial in *Larix leptolepis*: assessment 1964.
29. J.T. Miller 1965b Provenance trial in *Larix decidua*. Second assessment report. Age 6 years
64. J.T. Miller 1973 An extraction thinning in Larch provenance trials.
16. J.T. Miller 1964 Provenance trial of *Pinus nigra*: establishment report
30. J.T. Miller 1965c Provenance trial in *Pinus nigra*. First assessment report
68. M.D. Wilcox and J.T. Miller 1974 *Pinus nigra* provenance variation and selection in New Zealand (See same title Silvae Genetica 24 (5–6): 132–140).
D.J. Cown 1974: Physical properties of Corsican pine grown in NZ. NZJFS (4)

# Chapter 4
# Species and Provenance Testing of Western USA Species

**15. I.J. Thulin 1963 Provenance trials of *Pinus attenuata.* Establishment report.**

This was a range-wide collection of seed from plots of this species, established by R.Z Callahan of the Institute of Forest Genetics at Placerville, California. The provenances range from southern Oregon to south California, at altitudes mostly 2500–4500 ft. The seedlots are from two or more trees collected within a plot, located as specified, e.g. Lake 1 Mix is from several trees in the Lake County plot 1, and Lakes 1–5 were from five trees in the same plot 1. Differences between progeny from the same plot are thus within-provenance. There were two hybrid lots with *P. radiata*, FRI (Forest Research Institute) 914 and a control-pollinated lot of Eldorado 2-1 x *P. radiata* pollen mix. Open-pollinated seed of the same *P. attenuata* female parent is lot 43 (Thulin 1963).

Seed was sown at FRI in October 1959, part as a replicated trial, raised as 2/0, and the rest raised as 1+1 at FRI and Milton nurseries. Heights in the nursery trial were measured in 1961. Mean heights ranged from 7.9 to 23.0 in. (the control-pollinated hybrid was 25.3 in. at FRI nursery). There was a negative correlation of height with latitude of origin.

Trials were planted in 1961 at Naseby, Karioi, Kaingaroa and Whaka. Karioi and Kaingaroa were planted as 7 × 7 lattice squares, with Naseby and Whaka as randomised blocks with four reps (three at Whaka). There were 16 trees per plot (12 at Whaka) which were blanked in 1962 after poor survival.

The sophisticated design from Cochran and Cox 1950 p.361 plan 12.4 was the first attempt at small plots of two rows of eight or six trees per plot. There was excellent documentation of these trials, good maps and diagrams of layouts, etc. (Sweet & Bollman 1962). It is quite different from other designs of the same era.



C. J. A. Shelbourne, M. Carson, *Tree Breeding and Genetics in New Zealand*,
https://doi.org/10.1007/978-3-030-18460-5_4

**M.P. Bollman 1966: The height growth of a provenance trial in *Pinus attenuata* in Kaingaroa Forest Cpt 905.**
This paper reported heights of 48 seedlots, including one control-pollinated seedlot of *P. attenuata x P. radiata.* At the time of assessment, there was early spread of *Dothistroma* defoliation throughout the trial which necessitated its removal. The design was a randomised complete block with 4 replications of 16-tree plots of each seedlot. Eleven seedlots were based on 2–6 parent trees, with the remainder with a single parent. There was high survival throughout and seedlot mean heights varied from 9.1 ft. for the hybrid lot to 3.0 ft. for a southern Californian seedlot from 4000 ft. altitude. The more northern provenances generally grew better than the southern. Other *P. attenuata* seedlots were at the lower end of the distribution, though there was one suspect and very well-grown lot (6.9 ft.) that probably had some radiata male parentage (Bollman 1966).

This trial was destroyed soon after this assessment because it was a focus of *Dothistroma* infection. The same applied to trials at Karioi and Whaka. The trial at Naseby, outside the *Dothistroma* area, was preserved and was subject to further assessment of which no records can be found. The best trees, about 15 of them at Naseby, were selected and used as pollen parents in a trial of *P. attenuata x P. radiata* hybrids. These progenies were planted in the South Island and have shown superior growth, form and frost hardiness, on very cold sites, exceeding the performance of the radiata progenies (Dungey et al. 2011).

**M.H. Bannister. Notes on *Pinus muricata* and *P. attenuata***
*P. muricata*

This was a good round up of impressions from visits by the late Martin Bannister to Inverness Ridge, Monterey, Lompoc and San Vicente. Martin notes 'blue' from Trinidad Head and Fort Bragg, mixed 'blue' and 'green' at Annapolis, Mendocino County (Co.), and 'green' at Inverness Ridge (Marin Co.). He also notes bushy habit in Monterey and further south and on Santa Rosa and Santa Cruz islands. Tree form on Cedros Island he acknowledges may not be *P. muricata*. He also reports on trials at the 1961 experiment at Reids Pinch in ACT which has radiata provenances and some muricata seedlots.

*P. attenuata*

This is an excellent account of *P. attenuata* by the late Martin Bannister. Ecologically it is very different from radiata and muricata and is a rough, bushy, low tree emerging from sclerophyll scrub, often as widely scattered trees. It also forms at its best a well-formed tree (as in the NZ provenance trials) and is dominant over quite wide areas. Thus, like muricata, it has a wide geographic distribution, wide range of habitats and wide morphological variation. It is more frost resistant and drought resistant than radiata or muricata and hybridises easily with radiata and 'offers virtually unlimited scope for introducing new genes into a *P. radiata* breeding programme' (a bit over the top!).

**H.S. Dungey, C.B. Low, N.J. Ledgard & G.T. Stovold 2011: Alternatives to *Pinus radiata* in the New Zealand high country: early growth and survival of *P. radiata, P. attenuata* and their F1 hybrid. NZJFS 41: 61–69.**
Fifteen CP families of *Pinus attenuata* × *P. radiata* hybrids and their open-pollinated pure-species progenies were tested across three trial sites in the South Island of New Zealand. These hybrids were designed to combine the cold and snow resistance of *P. attenuata* with the faster growth of *P. radiata*. At ages 4 and 8, the hybrids were the tallest at the two semi-continental sites and had the most acceptable crop trees when compared with the pure species. At the mildest site, *P. radiata* was the tallest and *P. attenuata* was the shortest. After an exceptional snowfall, snow damage was recorded to be greatest in *P. radiata*, *P. attenuata* had the least, and the hybrids were intermediate between the two parent species. The *P. attenuata* × *P. radiata* hybrid also offers a real alternative to pine species that are prone to spreading in the New Zealand high country (e.g. *P. contorta*) (Dungey et al. 2011).

A corrigendum to this paper is available at http://www.scionresearch.com/__data/assets/pdf_file/0018/34344/NZJFS41201195_96_-DUNGEY.pdf

**J.W. Gilmour 1962 (uncertain): Dothistroma needle blight**
*Dothistroma* was first observed by the late John Gilmour on *P. attenuata x P. radiata* hybrids at NZFP in 1962 but identified only in 1964. There were lots of observations of this disease but not much on other pines at this stage. It was found on *P. ponderosa* but only on 20–30-year-old trees and caused severe defoliation.

**62. C.J.A. Shelbourne, J.A. Zabkiewicz and P.A. Allan 1973: Monoterpene composition in provenances of Pinus *muricata* planted in New Zealand.**
Two of possible three chemical races of *P. muricata* were found in a countrywide survey of *P. muricata* stands, with contrasting monoterpenes (from BH (breast height) sample). A provenance with predominant alpha-pinene contrasted with a provenance with delta-3-carene. Most of the planted area was of the delta-3-carene race that comes from either Monterey, Inverness or southern Sonoma County and is known as the green race. These are inferior in growth and form to the alpha-pinene 'blue' race of which only about 100 ha had been planted in Kaingaroa Forest versus 2400–2800 ha of the inferior 'green' provenance (Shelbourne et al. 1973).

This study was needed to identify the provenance of existing stands. *Pinus muricata* is distributed from latitude 42° near Eureka to 32° in Mexico. The largest continuous stands are between Fort Bragg and Fort Ross, but there is an abrupt change at latitude 38° 42′ from 'green' to 'blue' provenance. The third chemical race is found from San Luis Obispo southwards, which contains sabinene, some alpha-pinene and terpinolene.

**63. C.J.A. Shelbourne 1973: Recent investigations of wood properties and growth performance in 'blue strain' *Pinus muricata*. NZ J. For. 19 (1): 13–45 1974.**
'There is a need for an exotic tree species better adapted to colder, higher altitude sites and poorer soils than radiata but with similar growth rates and utilisation characteristics'. This was Shelbourne's rationale behind this interest in *P. muricata*, and

the limited evidence seemed to point favourably to this species. The paper reviewed the results of several comparative growth plots and utilisation studies of 'blue' alpha-pinene *P. muricata*, including wood properties (smaller pith to bark gradient, more summerwood-springwood contrast, more resin, better visual timber grades and higher stiffness). Heights were about 85% of radiata's and better than radiata on a few high-altitude sites (Shelbourne 1973).

My attitude to the species was probably too optimistic and interpreted the limited data too favourably. In the event it did not live up to expectations especially regarding frost resistance and growth on poor sites, but if radiata was threatened by major disease problems that muricata wasn't, it might be useful.

The monoterpene story (see BR 63) was the key in sorting out the differentiation of populations of *P. muricata*. The single plots of each provenance at Reids Pinch in the ACT were the only early provenance trial with *P. radiata*. Height of the Ft. Bragg provenance was the same as Monterey *P. radiata* at age 10 years.

Recent studies of wood properties, including unpublished work with a beta ray densitometer and a sawing study were reviewed. Data on comparative growth of blue *P. muricata* and *P. radiata* at Cpts. 1118 and 1119 Kaingaroa, at Cpt. 7 Waiotapu and at Golden Downs showed higher basal areas of *P. muricata*. It was difficult to distinguish relative growth rates because of stocking differences. Heights of *P. muricata* were always lower than radiata's by about 15%.

This was a comprehensive review of all available information. Where I went wrong was in hypothesising that *P. muricata* was better adapted on frosty and high-altitude colder sites. Also branching habit was such that it did not respond well to radiata's wide spacing regimes.

**40/6939 C.J.A. Shelbourne 1974: *Pinus muricata* provenance variation in height and state of flushing of the terminal bud, 18 months from sowing.**
Mean height of provenances showed an increase from Humboldt County in the north through Mendocino and Sonoma counties to Marin County in the south. Clinal increase in state of flushing followed the same north-south pattern with more flushing in the south. The NZ blue seedlots from Cpt. 1118 Kaingaroa apparently came from Mendocino County, and it was thought, erroneously, that the Cpt. 1054 green seedlot came from Sonoma. It was actually from Marin County (Shelbourne 1974).

**147/7052 C.J.A. Shelbourne 1976? Two-year heights of 35 wind-pollinated progenies of *P. muricate*.**
Family mean heights of 35 open-pollinated families of 'blue' muricata, planted in Cpt. 1038 and 885 Kaingaroa, and at Karioi, 2 years after planting showed large differences. Family x site interaction between the three frosty sites was not significant. Frost damage was severe at two of the sites (Shelbourne 1976).

**180/7091 M.D. Wilcox, C.B. Low and M.H. Bannister 1978: Assessment at age 6 years of the *Pinus muricata* gene pool experiment at Rotoehu forest.**
This experiment comprised 236 open-pollinated families of 16 provenances, about 15 families per provenance. At age 6 years Sonoma Co. 'green' was the best

grown population. 'Green' strain from Kaingaroa (later known to be Marin County) had healthy foliage and was of excellent form but slower growth. Kaingaroa 'blue' also had excellent form and good growth. The northern 'blue' from Trinidad (Humboldt County) was very slow-grown. Cedros radiata had the same good growth as the Sonoma green lot. The seedlots from Monterey southwards were variously inferior.

There were big differences in bole straightness and in needle retention, with the southern provenances from Santa Barbara being the worst. From later assessments the provenances from Monterey southwards were inferior in growth as well (Wilcox et al. 1978).

**R1312. C.J.A. Shelbourne, M.H. Bannister and M.D. Wilcox 1982: Early results of provenance studies on *Pinus muricata* in New Zealand. NZ J. For. 27(1): 50–66.**

This paper reports results of the 1972 Rotoehu trial of all provenances of *P. muricata* at age 6 years and the results of the 1973 trials of a more restricted group of provenances at Kaingaroa, and 10 countrywide sites, at age 5 years from planting.

In the Bannister range-wide study of 15 open-pollinated families per provenance, the Sonoma green provenance ranked easily top for volume and DBH. Trinidad (Humboldt County) was slow-grown, and Ft. Bragg (Mendocino County) and the Kaingaroa blue from Cpts. 1118 and 1217 ranked second as a group, but San Luis Obispo also ranked with them for volume. All provenances south of Marin County were poor for straightness and needle retention. Cedros Is. *P. radiata* was top equal with Sonoma in height and volume (Shelbourne et al. 1982).

The Bannister trial on a warm, low-altitude Bay of Plenty site showed that the southern provenances from Monterey southwards had poorer straightness than the provenances from further north. Sonoma was outstanding for growth and straightness, and the local 'blue' seedlots and the Mendocino provenance grew well (Bannister 1982). Local 'blue' from Kaingaroa had best straightness (check later internal report on subsequent assessment).

In the Shelbourne experiments at Kaingaroa, altitude 500 and 620 m, very flat and frosty (in retrospect the worst possible sites at Kaingaroa for such trials), various Sonoma seedlots, including some blue x green crosses, ranked first for height and diameter but 'best' was only 4 m tall at age 5 years. Growth of provenances which included Trinidad, Mendocino, Marin, Sonoma and two NZ blue (Mendocino) and one NZ green (Marin) was very slow on all sites except Kaingaroa 885, Kaweka, Golden Downs and Mahinapua. There Sonoma equalled or outgrew the blue NZ lots from Kaingaroa and Ashley seed stand. This was true over all sites, but the trials were too young for reliable assessment (Shelbourne 1982).

At Golden Downs and at Kakahu, in Canterbury, gene resource stands were planted later. A Sonoma seed stand was planted in Kaingaroa which grew well but got felled by mistake. Blue muricata showed up well for crown retention in the face of *Dothistroma*.

*P. muricata* did get planted in various places but didn't respond well to early thinning to waste (big branches) but did show up well for crown retention.

**24. I.J. Thulin, J.T. Miller 1966: Provenance trials of *Pinus contorta* planted from 1958–1961; Establishment report.**
Provenance trials in *P. contorta* were planted from 1958 to 1961 at 16 sites. This report includes nursery heights and survival 1 year after planting. The provenances include ssp. *contorta* (coastal low elevations), ssp. *bolanderi* (Mendocino County, California coast), ssp. *murrayana* (Oregon Cascades and Sierra Nevada) and ssp. *latifolia* and include the rest, i.e. all interior lodgepole in British Columbia and south of Washington, Oregon and Montana.

In the 1960–1961 trials of 28 seedlots, 14 were collected by Egon Larsen, 11 were from the US Forest Service and commercial seedlots and 3 were from NZ. Seven of the seedlots were from coastal ssp. *contorta* and ssp. *bolanderi*, and R569 came from Cpt 51 Kaingaroa and was possibly an intermediate *contorta x latifolia*. There were 13 ssp. *murrayana* lots, 5 ssp. *latifolia* and an intermediate ssp. *contorta x murrayana* and 1 *murrayana x latifolia*. One Golden Downs seedlot from Cpt. 56 was ssp. *contorta x latifolia*.

The 1960–1961 trials with 29 seedlots were planted at Whaka, Kaingaroa, Esk, Karioi, Gwavas, Golden Downs, Granville, Tara Hills, Naseby, Herbert, Earnscleugh and Rankleburn. All sites had three or four plots. Number of trees per plot was not given but varied from 20, 32 to 50 at Kaingaroa.

***Note*** Esk was planted with 25 plots (2 trees/plot) and Kaingaroa R. 320/3 with 25 plots (2 trees/plot), both 5 × 5 Latin square designs.

**27/6926 J.W. Hignett 1971: Provenance variation in height and wood density in *Pinus contorta.***
The complete *P. contorta* experiment has 28 provenances at 7 sites. Twenty of these at age 10 years were measured for height and wood density at two sites, Kaingaroa and Whaka. Seven coastal provenances, one ssp. *contorta x latifolia* and one *contorta x murrayana* failed to show significant differences in height and wood density. Plot sizes varied from 20, 32 to 50 with number of replications of 3–4 (Hignett 1971).

**118/7025 J.T. Miller 1976: Assessment of *Pinus contorta* provenance trials at Tara Hills.**
At age 5 years, the two transitional seedlots are best for growth and health. The inland provenances of ssp. *latifolia* from BC were a little slower-grown but suffered some damage from drought. ssp. *murrayana* grew more slowly but remained healthy.

**38. J.T. Miller 1968: The genetic improvement of *Pinus contorta* in New Zealand.**
This was a similar but less exhaustive review of *P. contorta* to the Douglas-fir one. However, it was much more oriented to genetic differentiation, as seen in provenance trial results at age 5 years and in past experience of provenance variation with different NZ seed sources (Miller 1968).

Three sections address the need for improvement, prospects in NZ and methods of improvement in NZ, with specific proposals. These involved plus-tree selection in two populations: Cpt. 51 Kaingaroa strain (probably *P. contorta x latifolia*) and Waiotapu Cpt. 2, probably Californian coast. They utilise open-pollinated progeny tests, convertible to open-pollinated seedling orchards, and clonal orchards, based on reselection on progeny test results among 100 selections in each population.

**55. J.T. Miller 1971: Provenance variation in growth rate and other characters in 6-year-old *Pinus contorta*. (See also establishment report 24. Thulin and Miller 1966)**

Provenance trials in *P. contorta* were planted at 16 sites from 1958 to 1961. The establishment report covered nursery heights and survival 1 year after planting. Seven seedlots were from coastal ssp. *contorta, bolanderi* and R569 from Cpt. 51 Kaingaroa. There were 13 ssp. *murrayana* lots, 5 ssp. *latifolia* lots and single intermediate lots of *contorta/murrayana, murrayana/latifolia* and *contorta/latifolia.*

The design and number of trees per plot are not given, but there were three to four plots of each seedlot at each site. 100 trees per plot or more would be expected based on current practice. At Esk and Kaingaroa, there were 25 replications and 2 trees per plot in 5 × 5 Latin square designs, a new innovation, possibly by Geoff Sweet.

The 1960–1961 trials of 29 seedlots were planted at 14 sites. Overall mean height of coastal lots at age 6 years was 8.3 ft.; the transitional group, 6.5 ft.; and inland group, 3.6 ft. Provenance heights were generally correlated with altitude of seed source. Stem lean was greatest for tall coastal provenances. The NZ Waiotapu strain was slightly shorter than Langlois, Long Beach and Seal Rocks provenances, and the Kaingaroa Cpt. 51 seedlot was the slowest-grown of the ssp. *contorta* lots at 7.45 ft. versus 7.9–9.1 for the other coastal seedlots. All the ssp. *murrayana* and *latifolia* lots were much slower-grown than ssp. *contorta* (Miller 1971).

***Comment*** The large number of sites (16) meant that precision of overall seedlot means was good in spite of significant provenance x site interaction. The very wide variation in height growth in the trials because of inclusion of ssp. *latifolia* and *murrayana* provenances probably contributed to provenance x site interaction, as did the wide differences in site environment. This study was a fine basis for an improvement programme, well planned and wide coverage of sites and seedlots. However, inclusion of the interior ssp. *murrayana* and *latifolia* provenances could well have been omitted.

The choice of the Waiotapu and Cpt 51 strains as two separate populations to breed from is questionable. Although there were no other mature populations to select from in NZ, probably a single breeding population including all available coastal provenances, including those in the trials, would have given more genetic gain at lower cost. The inclusion of only five native coastal provenances required introduction of more coastal provenances, but this occurred too late to be of application, and sadly the breeding programme was soon to be abandoned for other reasons.

A good feature of wind-pollinated conifer populations seems to be they do not suffer much neighbourhood inbreeding, so that open-pollinated family testing is more reliable than in insect-pollinated species like eucalypts. The question (as faced also in the eucalypt species and Douglas-fir) was whether to rely on existing NZ populations to select and breed from or a range of native provenances.

**R. 993. C.J.A. Shelbourne and J.T. Miller 1976: Provenance variation in *Pinus contorta*. 6-year results from IUFRO seedlots in New Zealand. IUFRO XVI IUFRO World Congress Norway 1976. Proc. Div. II:140–145.**
Fifteen seedlots including five NZ lots were assessed at Cpt 905 Kaingaroa (alt. 530 m) and Karioi (alt. 930 m) at age 4 years from planting. The northernmost lot from Gold River in Vancouver Is. grew second slowest, but the lot from Queets, Washington, which was originally growing on muskeg, was slower-grown and apparently was a representative of the northern 'muskeg' race, found on coastal sites from Vancouver north to Alaska (my experience with this provenance was with the UK Forestry Commission).

This is different from the 'shore pine' which is represented by the other more southern coastal seedlots of the IUFRO seed collection.

At Kaingaroa, apart from the Queets and Gold River and Vancouver Is. lots, there were only minimal differences in height, but diameter at 10 cm above ground increased substantially with decreasing latitude, with the biggest from Manchester, California, latitude 38° 97′. The Port Orford and Gold River lots appear to be transitional types to the interior lodgepole pine (ssp. *latifolia*). The poor growth of the Ft. Bragg lot from California may be because it also came from a muskeg peat site and seemed to be different genetically from the shore pine seedlots.

At Karioi, heavy possum browsing only affected the US provenances from California (Samoa, Ft. Bragg and Manchester) and was very heavy on four NZ lots of the Waiotapu strain. There was almost none on the Karioi seedlot, the so-called Kaingaroa strain, probably an intermediate ssp. *contorta/latifolia* to interior lodgepole. It was interesting how specific the possums were with regard to provenance (Shelbourne & Miller 1976).

***Comment*** Reported in 1976, assessed in 1975, these results preceded the eventual abandonment of the species for breeding and of much further plantation use. This was due to the development of techniques of establishing *P. radiata* on frosty sites, combined with the bad weed potential of *P. contorta*.

**192/7118 O. Mohrdiek and J.T. Miller 1980: Five-year assessment of open-pollinated progeny tests in *Pinus contorta*.**
The assessment of open-pollinated tests of 100 families each of Waiotapu and Kaingaroa strains was done on 2 sites in the NI and 2 in the SI with the objective of backward selecting the best parents for use in clonal seed orchards. Heavy opossum damage occurred on the Waiotapu strain versus almost none on the Kaingaroa strain. This was faster growing on three of the four sites, but Waiotapu was still the fastest at Kaingaroa.

It is recommended that the Kaingaroa strain should be favoured over the Waiotapu. However, the breeding programme was abandoned in the meantime, and these progeny trials are unlikely to be utilised (Mohrdiek & Miller 1980).

**9. G.B. Sweet and M.P. Bollman 1962: Provenance trial in *Pinus ponderosa*: A combined establishment and first assessment report.**
This was a preliminary trial of five Californian provenances from latitude 38° 30′, altitude 3000 ft., planted at seven sites in 1955. All five seedlots were located close together in California, with a maximum separation of 50 miles. Previous work was a unique trial of 13 provenances, planted in 1929, which included three Californian provenances from west of the Sierra Nevada, one east of the Sierra, six from BC between Coast Range and Rockies and three from east of the Rockies in Colorado and New Mexico. An assessment by A.M. Moore and I.J. Thulin showed superiority of Californian lots.

As a preliminary trial, five seedlots from 2600–3700 ft. altitude were sown at FRI in 1952 and lined out as 1+1 at FRI and Milton nurseries. Plots were planted at 8 × 8 ft. spacing, varying from one plot per seedlot to four at Kaingaroa, Gwavas, Golden Downs, Berwick and Naseby. Plot size was varied from 289 to 324 trees at Kaingaroa, Gwavas, Golden Downs, Berwick and Naseby. Survivals at 1 year varied from 67% at Berwick to 91% at Kaingaroa and 70–78% among seedlots.

The assessment of height and survival was done at age 7 years (1962). Survival averaged over all sites was 58 and was 95–99% at Kaingaroa and other sites. Overall means for height ranged from 7.45 ft. (seedlot 137, altitude 3700 ft) to 7.88 ft. (seedlot 138, altitude 2600 ft.). This was based on a sample of 50 trees/plot (Sweet & Bollman 1962).

***Comment*** Genetic differences among these five seedlots (1–4 plots per seedlot per site) were minor and one has to ask: 'Why do such a trial on so many sites, with so many trees per plot and variable and scanty replication?' The precision of seedlot comparisons at each site was poor, but the quality of field work and the analysis and write-up were good.

***Reference*** A.M. Moore 1944: *Pinus ponderosia*: a comparison of various types grown experimentally at Kaingaroa SF. N.Z. J. For. Vol 5 no. 1.

**40. I.J. Thulin and M.D. Wilcox 1970: Ponderosa pine provenance trials (1960–61). Establishment report.**
This report describes the subdivision of *P. ponderosa* var. *ponderosa* into the Californian type and the North Plateau type and *P. ponderosa* var. *scopulorum* into the Southern Interior, Central Interior and Northern Interior type. It also lists *P. ponderosa* var. *arizonica*, *P. washoensis* and *P. jeffreyi*. There is a good review of the Moore 1944 Wairapoukao trial (unreplicated). They also review the 1928 Oregon-Washington study and California study of 1933–1938. The history of NZ planting and seedlots is valuable, with heights in the Moore trial of the Californian provenance at Kaingaroa of 99 ft., BC of 63 ft. and var. *scopulorum* of 42 ft. 77,504 acres of all provenances was established, 66,000 acres in the NI, 3000 acres of good Californian and the remainder var. *scopulorum*.

Eight trials were established in 1960, eight in 1961 and one in 1962. The design was randomised blocks, up to four reps per site. Plots of 32 trees were planted at 8 × 8 ft.

spacing. Lists of all *P. ponderosa* seedlots in NZ and weights of seed of each were given: 64 seedlots in tests, including 8 *P. jeffreyi*, 1 *P. washoensis*, 38 Californian, 14 North Plateau, 1 var. *scopulorum*, 1 ssp. *arizonica*, 1 ex Naseby Cpt. 1 and 1 ex Queenstown. There were 14 sites with 3 plots, 13 sites with 4 plots and Esk with 1 plot/provenance. Nursery heights for each year and site descriptions and seedlot descriptions were included.

***Reference*** A.M. Moore 1944: *Pinus ponderosa*: a comparison of various types grown experimentally at Kaingaroa SF. N.Z. J. For. Vol 5 no. 1.

### 47. M.D. Wilcox 1970: Ponderosa pine provenance trials (1960–61) assessment six years after planting (1967).

Heights were measured on 16 trees out of 32 per plot at most sites. Height and survival of 64 provenances at 16 sites were assessed 6 or 7 years after planting. No important differences in survival were detected. Two provenances of 'Californian type' grew faster than most at nearly all locations. These were provenance 666 Canyonville, Rogue River NF, OR from 1000 ft. altitude (in SW Oregon), and provenance 667 Grants Pass, OR, also from 1000 ft. altitude, was another winner. At Gwavas they were 12 ft. tall at 7 years from planting. Half the 40 provenances, widely tested, ranked in the top 25% at one or more locations.

On most NZ sites North Plateau provenances grew slower than Californian type but were among the fastest in inland Otago. Conversely, high-ranking Californian-type provenances in the NI and Nelson were low ranked in Canterbury and Otago.

A more detailed assessment was done at Cpt. 905 Kaingaroa. Many different analyses were done of various fractions of the sites and provenances. Special attempts were made to discover the source of the provenance x site interaction. Certain provenances were shown to contribute most of the interaction. Height, height increment, needle length and branch number per whorl at Cpt. 905 Kaingaroa all showed significant differences.

The main conclusions were that two 'Californian-type' provenances, 666 and 667, grew best at nearly all sites; that certain North Plateau provenances grew best in dry, hot/cold Otago climates. *P. jeffreyi*, *P. ponderosa* var. *scopulorum*, *P. washoensis* and *P. ponderosa* var. *arizonica* were markedly inferior, and Naseby seedlots of *P. ponderosa* and *P. jeffreyi* were very slow-grown. Vigorous Californian provenances suffered little from frost.

***Comment*** Sadly, the impact of *Dothistroma* on all seedlots of *P. ponderosa* was disastrous, except in Otago and other SI areas. The only place for this species appeared to be in Otago, but wood and sawing properties were not good.

### R.D. Burdon and C.B. Low 1991: Performance of *Pinus ponderosa* and *Pinus jeffreyi* provenances in New Zealand. Can. J. For. Res. 21:1401–1414.

This was a final assessment in 1984 at age 22–24 years of two sites in the SI of NZ where *P. ponderosa* provenance trials had escaped *Dothistroma*. References were not listed of the previous Branch Reports by Sweet, Thulin and Wilcox (1970) and Wilcox (1970) (age 6 years) though they were referred to as 'unpublished'. The long Introduction however is an excellent account of 'pondy' in NZ, from the 1920s onwards.

This assessment was based mainly on Hanmer and Tara Hills, with *Dothistroma* results from Gwavas and Kaingaroa. Thus, the result applies only to semi-continental sites in SI. *Dothistroma* has decimated even the best provenances, where it is present. The North Plateau provenances did well at the remarkably dry site of Tara Hills, whereas the low-altitude Sierra Nevada and Coastal provenances did best at Hanmer. Earlier *Dothistroma* data from Gwavas and Kaingaroa indicated that North Plateau provenances were better (but not good enough).

**R. 2115. J.T. Miller and C.J.A. Shelbourne 1984: Sitka spruce provenance trials in New Zealand. IUFRO Working Party S2.02.12 on Sitka spruce provenances. Edinburgh, UK. Sept. 1984.**
There were two series of trials, planted in 1960 and in 1972, which were assessed at ages 6–8 and 23–25 years, and for the 1972 trials, at ages 3 and 12 years. The 1972 series contained 11 provenances from Washington, Oregon and California from Alan Fletcher's IUFRO collection and also 3 commercial lots from the USA and NZ Conical Hill. The field design for the 1972 trials was randomised complete blocks with 5–10 replications of 10-tree row plots at 4 sites in each island (8 in all).

Assessments at four sites, Hunua, Karioi, Slopedown and Whaka, were made for diameter, height and aphid defoliation score at age 12 years (1 = zero defoliation). The seedlots had a near-complete latitudinal spread from Vancouver Is. to northern California. Diameter was strongly inversely correlated with latitude of origin. The commercial lots followed the same latitudinal trend. Aphid attack was heavy, except at Hunua. Nothing can apparently improve the poor standing of Sitka spruce in NZ with regard to aphid attack. This has occurred everywhere, and this species' sensitivity has been acute. There is some improvement in growth and health when planted next to pines or other species. Interestingly, NZ Conical Hill is the slowest-grown lot, implying a far northern source, perhaps Queen Charlotte Islands.

The 1960 trials contained four seedlots collected by Egon Larsen, including Big Lagoon and Caspar, California; Waldport, OR.; Conical Hill, NZ; and Vancouver Is. (island), and a commercial lot from Naselle, Washington. The Vancouver Is. lot was only planted at Golden Downs and Granville, where it grew by far the slowest at age 6–8 years. The commercial Naselle, Washington, lot grew slowest at Golden Downs. Replication was very sparse, not shown, but was probably one to three large plots per site, thus resulting in very poor precision. Significant differences in provenance means at age 6–8 years were only shown in height at Waitangi, Glenbervie, and just at Woodhill and Hunua, and not at the other eight sites. Overall sites, Big Lagoon and Caspar, California, grew better than Waldport, Oregon, and Conical Hill, NZ.

At age 23–25 years, only Conical Hill, NZ, could be shown to grow slower than the other provenances at Hunua, Esk and Whaka. By age 24, 11 of the original 17 trials had been abandoned because of poor growth and survival with only 3 trials worth assessment. Predominant mean heights were only up to 15 m at Whaka and the Californian provenances were best. This has to be regarded as a failed species on the basis of these trials. There is a suggestion that part of the problem is mycorrhizal and part, aphid. The species does not live up to its reputation from the UK though of course there, it is growing at much higher and colder latitudes, the main provenance being Queen Charlotte Islands, from north of Vancouver Is. Neither does it match its amazing growth in the Olympic Peninsula or the growth of Douglas-fir in NZ. One has to admit that this species' associates, *Thuja plicata* and *Tsuga heterophylla*, also do not match Douglas-fir's performance in NZ (Miller & Shelbourne 1984).

**R. 2621. R.D. Burdon and J.T. Miller 1995: Alternative species revisited: categorisation and issues for strategy and research. NZ Forestry Aug. 1995.** Three main categories of species can be distinguished, alternative to *P. radiata*:

1. Special-purpose species with utilisation unsuited to *P. radiata* and capable of producing anything from high-value solid wood to pulpwood and firewood
2. Extreme-site species which can grow well on sites where radiata can't
3. Contingency species which could replace radiata if disease or insect attack (or climatic change?) rendered radiata a write-off (insect attack or a new fungal disease seems most likely causes, to which any replacement would have to be resistant or tolerant)

This categorisation is just one for the alternative species. It doesn't break down the special purpose species into more useful categories, which have real products associated. At the time of writing (1995), there was increasing international and domestic concern about biodiversity and sustainability. Radiata was always vulnerable to criticism and to pest attack on these grounds, yet its 'front-line genetic defences' rested on maintaining its genetic variability. FRST had backed research on alternative species since 1993 and earlier. However, work in these areas was much governed by industry interest in short-fibred hardwood pulp and structural timber, respectively, through the Eucalypt Breeding Cooperative and the Douglas-fir Cooperative. Inevitably, any 'alternative species' in any category was judged based on its comparison with radiata in growth, ease of establishment and adaptability to different environments.

***Comment*** Species are listed by the three categories and pretty much correspond to their position in 2014. The eucalypts have changed: *E. saligna* has gone, *E. nitens* is unsuited to the warmer sites in NI but good in the southern SI, and *E. regnans* is not much planted any more, neither is *E. delegatensis*. None of the other pines would really be given much value as contingency spp., except perhaps *P. muricata*, yet it is too closely related to *P. radiata*. *P. taeda* should never have got a mention in view of its hopeless wood properties. *Juglans* and *Paulownia* have fallen by the wayside, and there is no demand for poplars for sawtimber.

Generally, the list seems to have shrunk a lot, based partly on ongoing research and practical experience. Douglas-fir has strengthened its position as a real alternative to *P. radiata* on some cold and snow-prone sites. *E. fastigata* is stronger as a short-fibred pulp and sawn timber species than previously believed. *C. macrocarpa* has been badly affected by disease. *P. contorta* is badly attacked by possums, at least in the faster-grown southern coastal provenances, and is an outcast on 'weed' grounds. Redwood has got more use through Soper-Wheeler planting.

There is a need to tie in with Wilcox 1993 NZ Forestry 38(3) article, 'Priorities for research on alternative tree species for wood production in NZ'.

**IR 83 PROPPS 718 J.T. Miller 1978: Species provenance trials of Sierra Redwood: establishment and first assessment.**
The Sierra redwood, *Sequoiadendron giganteum*, has been planted as a single tree or in small groups or avenues, but there are no known plantations in NZ. Lauren Fins, at the time a graduate student of Bill Libby's at Berkeley, UC (University of California), made an extensive seed collection of this species, and seeds from 16 groves were donated to NZ.

Seed was bulked by groves. Seeds of 16 lots plus 1 from NZ Raincliff Forest were sown in 1975. Cuttings were taken of seedlots that were short of stock in 1977. Seedlings were planted in 1977 and cuttings grown on for planting 1 year later in 1978. Stock was planted at six sites, Hanmer Cpt. 1, Beaumont Cpt. 3, Lyndon Hill and Craigieburn Forest Park 1977 and, in 1978, Beaumont Cpt. 3 (extension), Kakahu Forest and Rai Forest. The field design was randomised single trees with variable numbers of seedlings/cuttings per seedlot. Spacing was 5 × 5 m with inter-planted larch to make up stocking. Survival and height were measured during releasing. Heights were 187–227 cm at age 5 at Beaumont, the best site.

**PDF 21243 in SIDNEY same number. W.J. Libby 1999: Observations of Giant Sequoia on South Island New Zealand.**
Little is known about giant sequoia in plantations. The majority of *Sequoiadendron giganteum* plantations more than 20 years old are in the Rhine Valley of Germany. France is now planting modest areas at altitudes over 300 m. Government and private forests in California are now planting 15% giant sequoia on better sites at 1000–2000 m. There is no established market for the timber, though it appears similar to coast redwood, *Sequoia sempervirens*.

A remarkable stand is the 80-year-old stand at La Belle Etoile in Belgium with an MAI of 38–48 $m^3$/ha/year versus Douglas-fir with 18 $m^3$/ha/year and Scots pine and Norway spruce with 4–8 $m^3$/ha/year.

## References


15. I.J. Thulin 1963 Provenance trials of *Pinus attenuata*. Establishment report

M.P. Bollman 1966: The height growth of a provenance trial in *Pinus attenuata* in Kaingaroa Forest Cpt 905.

H.S. Dungey, C.B. Low, N.J. Ledgard & G.T. Stovold 2011: Alternatives to *Pinus radiata* in the New Zealand high country: early growth and survival of *P. radiata, P. attenuata* and their F1 hybrid. NZJFS 41:61-69

62. C.J.A. Shelbourne, J.A. Zabkiewicz and P.A. Allan 1973: Monoterpene composition in provenances of Pinus *muricata* planted in New Zealand.

63. C.J.A. Shelbourne 1973: Recent investigations of wood properties and growth performance in 'blue strain' *Pinus muricata*. NZ J. For. 19(1):13-45 1974

40/6939 C.J.A. Shelbourne 1974: *Pinus muricata* provenance variation in height and state of flushing of the terminal bud, 18 months from sowing

147/7052 C.J.A. Shelbourne 1976? Two year heights of 35 wind-pollinated progenies of *P. muricata*

180/7091 M.D. Wilcox, C.B. Low and M.H. Bannister 1978: Assessment at age 6 years of the Pinus muricata gene pool experiment at Rotoehu forest.

R1312. C.J.A. Shelbourne, M.H. Bannister and M.D. Wilcox 1982: Early results of provenance studies on *Pinus muricata* in New Zealand. NZ J. For. 27(1): 50-66.

24. I.J. Thulin, J.T. Miller 1966: Provenance trials of *Pinus contorta* planted from 1958-1961; Establishment report.

27/6926 J.W. Hignett 1971: Provenance variation in height and wood density in *Pinus contorta*

118/7025 J.T. Miller 1976: Assessment of Pinus contorta provenance trials at Tara Hills.

38. J.T. Miller 1968: The genetic improvement of *Pinus contorta* in New Zealand.

55. J.T. Miller 1971: Provenance variation in growth rate and other characters in 6-year-old *Pinus contorta*.

R. 993. C.J.A. Shelbourne and J.T. Miller 1976: Provenance variation in *Pinus contorta*. 6-year results from IUFRO seedlots in New Zealand. IUFRO XVI IUFRO World Congress Norway 1976. Proc. Div. II:140-145.

192/7118 O. Mohrdiek and J.T. Miller 1980: Five year assessment of open-pollinated progeny tests in *Pinus contorta*.

9. G.B. Sweet and M.P. Bollman 1962: Provenance trial in Pinus ponderosa: A combined establishment and first assessment report.

40. I.J. Thulin and M.D. Wilcox 1970: Ponderosa pine provenance trials (1960-61). Establishment report

47. M.D. Wilcox 1970: Ponderosa pine provenance trials (1960-61) assessment six years after planting (1967).

R.D. Burdon and C.B. Low 1991: Performance of *Pinus ponderosa* and *Pinus jeffreyi* provenances in New Zealand. Can. J. For. Res. 21:1401-1414.

R. 2115. J.T. Miller and C.J.A. Shelbourne 1984: Sitka spruce provenance trials in New Zealand

R. 2621. R.D. Burdon and J.T. Miller 1995: Alternative species revisited: categorisation and issues for strategy and research. NZ Forestry Aug. 1995.

IR 83 PROPPS 718 J.T. Miller 1978: Species provenance trials of Sierra Redwood: establishment and first assessment.

PDF 21243 in SIDNEY same number. W.J. Libby 1999: Observations of Giant Sequoia on South Island New Zealand.J.M. Fielding 1961: Provenances of Monterey and Bishop pines. For. Tim. Bur. Aust. No 38.

Wilcox (1993) NZ Forestry 38 (3) article, 'Priorities for research on alternative tree species for wood production in NZ'. Aust. Bull. No. 11.

# Chapter 5
# Douglas-Fir Provenance and Breeding

**14. G.B. Sweet 1963: Some provenance differences in *Pseudotsuga menziesii*. 1. Seed characteristics.**

The relationships of photoperiod, germination temperature, degree of stratification, germination %, provenance and seed weight were analysed with data from several conifers including other Douglas-fir. The 1956 collection of 30 coastal provenances by Egon Larsen is the material studied in a mainly physiological study. Data on seed weight and seedlot formed the basis. Correlations of seed weight with altitude were 0.76, with frost-free days −0.57 and with latitude, −0.31. The report gives information about all seedlots. App. 1. 1000 seed wts., App. 3. germination capacity, App. 4. Germination after 42 days of all seedlots.

Twenty-three provenances from Larsen's collection of coastal provenances, with two NZ seedlots, were sown in an RCB design with three replications in FRI nursery (after planting the field trials of 1959).

Heights after first and second year from sowing and phenological data were recorded. Heights were correlated with mean temperature of the coldest month of origin. Order of bud burst differed among provenances but was, apparently, unrelated to climate of seed source. Lateral buds burst before the terminals and the time lag was closely correlated with the temperature of the seed source. There were no apparent differences between provenances in number of growth flushes, but there were some weak correlations of bud set and length of frost-free growing season.

***Comment*** There was a big range in 1- and 2-year provenance mean heights, 5.7–30.2 in., but some of this was due to altitude and continentality. Darrington, Washington, was 13.4 in. versus 654 Caspar, California, of 24.5 in. R. 530 was 19.0, the Kaingaroa control. Unfortunately, these provenances mean heights were never related to the Kaingaroa seedlot in future plus-tree selection.



C. J. A. Shelbourne, M. Carson, *Tree Breeding and Genetics in New Zealand*,
https://doi.org/10.1007/978-3-030-18460-5_5

**21. G.B. Sweet 1964: The establishment of provenance trials in Douglas fir (*Pseudotsuga menziesii*) in New Zealand.**
This is a monumental report of some 65 pages covering a total of 83 provenances in 2 series, each containing different provenances, the first planted in 1957 and the second planted in 1959. The actual text in the report only occupies 20 pages, and aspects of field design are hard to extract.

The first series was of commercial seed collections, not coastal, from latitudes mainly 48° in Northern Washington, e.g. Darrington (48° 10′) to 42° 30′ at Siskiyou NF in southern Oregon. Information on seed and nursery performance repeats much of previous reports especially of 1959 series.

The second series was of 30 provenances, collected from known locations by Egon Larsen in 1956. The text only occupies 20 pages and aspects of field design are hard to extract.

1957 series: 7 sites, 100–196 trees/plot, 1–3 plots per site. Rotoehu, Kaingaroa, Patanamu, Golden Downs, Hanmer, Berwick.

1959 series: 20 sites, 121–144 trees/plots, 1–3 mainly 1 plot per site. Waitangi, Glenbervie, Woodhill, Ahuroa (Jenkins) Warkworth (P. Wech), Whangapoua, Maramarua, Gordonton (A. Gower), Rotoehu, Rapanui (NZFP), Kaingaroa, Kaingaroa Cpt. 634, Patanamu, Gwavas, Golden Downs, Hanmer, Granville, Rankleburn, Rankleburn John O'Groats Rd.

Appendix 1 contains full information of seedlots for each series, with detailed meteorological data from nearest provenance station for 1959 series; Appendix 2 seed data, seed and replicates; Appendix 31,957 series site details with experiment maps for each site; Appendix 4 for 1959 series the same; Appendix 51,957 series survival and stocking of plots after blanking, as summarised above, seemingly the only source of design information; and Appendix 6 for 1959 series. Nursery heights, at 1 year at FRI and at 2–3 years at FRI and Milton for 1957 series based on 2-year height on altitude of seed source for both series.

***Note*** This report is accessible as GTI Branch Report 21 in the Scion library.

**23. G.B. Sweet 1964: The assessment six years after planting of a provenance trial in Douglas fir (1957 series).**
The seedlots involved in these trials were mainly commercial lots from west of the Cascades in BC, Washington and Oregon. They were not collected from coastal populations in general, and there were none from California except from the Sierra Nevada.

Latitudes were mainly from 48° in northern Washington, e.g. Darrington (48° 10′) to 42° 30′ at Siskiyou NF in southern Oregon. There were a few seedlots from BC Vancouver Is. and one from Prince George latitude 54°.

Survival, height and stem straightness (three sites only) were assessed on 50 trees per plot. Analysis was of plot means for provenances, trial areas and interaction pooled with residual. Provenance effects were highly significant for height and just significant for survival, and site × provenance interaction was not significant.

Provenance mean heights varied from 2 ft. to 9.6 ft. NZ lots averaged 8.7, 8.8 and 9.1 ft. Best growth was from Snohomish Co. (Darrington and Granite Falls), Snoqualmie NF and Olympic Peninsula.

There was little attempt to interpret these results except in looking at correlations with altitude and latitude of seed source for height and survival. The four NZ seedlots ranked within the first 13, and their heights were >8.7 ft. relative to Darrington with 8.9 ft. Height was correlated with survival and straightness. There was no site × provenance interaction for height or survival.

**G.B. Sweet 1965: Provenance differences in Pacific coast Douglas fir. Silvae Genetica 14(2): 46–56.**

Heights and phenology of the 1959 Larsen seed collection (plus some Ching and Beaver seedlots from Corvallis) were assessed over 2 years in a three-replication layout in FRI nursery (initiated after planting of provenance trials) (see BR 18). Heights of provenances were significantly related to several climatic features, best with mean temperature of coldest month, 0.79. The time lag between lateral bud and terminal bud burst varied among provenances and was correlated with temperature of coldest month. Provenances from milder climates at the coast had longer time lags.

***Comment*** Objectives and methods of this early study were largely predictive, based on nursery behaviour, to allow checking of source of seed. Statistics were regression and correlation of bud burst with climate of source, also, provenance height with climate. The replicated study of 1- and 2-year heights revealed big differences among coastal provenances.

Mean heights at age 2, as given in GTI Branch Report 18:

631, **13.4**, 632, **14.1**, 634, **17.8**, 636, Deadwood, Ore.; **18.0**, 638, **14.0**, 639, **15.5**, 640, **17.4**, 641, Fourmile, Ore.; **21.1**, 642 Berteleda, Cal.; **21.6**, 643, **5.7**, 644, **11.8**, 645, **15.9**, 646, Six Rivers, Cal.; **19.2**, 647 Mad River, Cal.; **20.3**, 648 Miranda, Cal.; **22.3**, 649 DeHaven, Cal.; **24.8**, 650, **18.6**, 652, **13.0**, 654 Jackson SF; **24.5**, 656, **16.5**, 657, **17.3**, 658 Stewarts Point, Cal.; **24.5**, 660 Santa Cruz, Cal.; **30.2** R 530, Kaingaroa, NZ, **19.0**, 538, **19.0**

Leading provenances at age 2 were:

Deadwood, Oregon, **18.0** in.; Fourmile, Oregon (Bandon), **21.1**; Berteleda, California (Crescent City), **21.6**; Six Rivers, California, **19.2**; Mad River, California, **20.3**; Miranda, California, **22.3**; DeHaven, California, **24.8**; Mendocino, California, **18.6**; Jackson SF (Ft. Bragg, probably Caspar), **24.5**; Stewarts Point, California, **24.5**; NW of Santa Cruz, California, **30.2** in. (note Stinson Beach was missing); Kaingaroa (R. 530?), **19.0**; FRI grounds, **19.0**

***Note*** Darrington, Washington, was only **13.4** in.; Siuslaw (Corvallis), Oregon, **17.8**; Wind River, Washington, **14.1**; and Siskiyou NF, Oregon, **17.4** (unfortunately Sweet did not give original seedlot numbers).

### Age 13 Years: DBH

These best provenances gave the following 13-year DBHs (as provided by Mike Wilcox). Last in brackets is the three-site analysis (all in cm.)

Deadwood, Oregon, 636: 10.1, 13.8 (**13.5**); Fourmile, Oregon, 641: 10.1 13.9 (**13.5**); Berteleda, California, 642: 10.6 14.4 (**14.0**); Six Rivers, California, 646: 8.7 13.0 (–); Mad River, California, 647: 10.8 13.8 (**13.5**); Miranda, California, Rossy's Ranch 648: 9.6 13.7 (–); DeHaven, California, 649: 9.9 13.5 (**12.9**); Jackson SF, Ft. Bragg, Caspar, California 654: 9.9 13.3 (**13.3**); Stewarts Point, California, 658: 9.9 13.5 (**13.1**); Santa Cruz, California, 660: 10.7 14.8 (**14.0**)

There were only 8 or possibly 9 lots common to the best 11 in the Sweet study and those in the 3-site rankings (first) of Wilcox. DBHs are consistently (12.9–14.8 cm) highest in the Wilcox 13-year means. If one excludes 646 Six Rivers, California (possibly not a 'coastal'), DBHs in the 6-site study all exceed 13.5 which is exceeded by the highest 12 out of 45 seedlots in the 3-site study.

***Comment*** Overall, the design of the field trials really prevents precise estimation and comparison of performance of provenances and particularly of provenance performance at a given site. Cutting down the number of sites to 3 allowed comparison of 45 seedlots but sometimes reduced precision with only 3 plots in total of some provenances. With 6 sites included, balanced comparisons of only 15 provenances were possible. This is a result of 'using up' so many plants, with 144 trees/plot over as many as 19 sites.

Mike Wilcox at this stage apparently did not think of selection from the 1959 provenances for his breeding population because that had been predetermined by plus-tree selection in the Kaingaroa (Washington) population.

### 37. M.D. Wilcox 1968: The genetic improvement of Douglas fir in New Zealand.

*Comment* This is a monumental review of literature about Douglas-fir (made in 1967–1968) with over 220 references. Much is devoted to practices and problems of growing, processing and using Douglas-fir, with less than one page on methods of genetic improvement. In his proposals for a breeding programme, selection of plus trees and establishment of seed orchards are central, with progeny testing also proposed.

The proposed programme is entirely operational. It utilises open-pollinated progeny tests based on selection in mature Douglas-fir stands in NZ with no attention to provenance or the large estate of provenance trials that were then 9 and 11 years old. The report is very much focused on seed production, seed itself, seed stands, seed imports, past and future. There was some thought about traits for selection of plus trees, with wood density mentioned and spiral grain. There was nothing about other traits. Variation in stem sinuosity, density, frost resistance, *Phaeocryptopus* resistance and strength of knotty timber received little attention. Provenance was accorded one line; 'Continued study of the variation due to provenance in the 1957 and 1959 provenance trials'.

With the wisdom of hindsight, this review should itself have been reviewed by colleagues and focused much more on how to plan the breeding programme. In spite of reference to a report, *Tree Breeding Methods* by Tony Shelbourne, this material did not get any consideration. This is strange because at this time there was much thought given to the future development of the radiata breeding programme by me and Rowland Burdon.

The importance of provenance was recognised, even that coastal Californian provenances grow fastest, but this was dismissed as an early trait that was not a good guide to later performance. In spite of good correlation of nursery heights with early results of the 1959 trials, it was concluded that mature straightness, branching, volume production, fast early height growth and wood density were the most important traits.

In retrospect, the task of initiating a new breeding programme should not have been loaded onto a new forestry graduate with no training in genetics and breeding. Furthermore, Ib was unwise in not getting more input from others on the structuring of a new programme. At the time, 1967, that Mike W. started this review, thinking about breeding strategy, breeding populations, progeny test designs, etc. was developing fast.

**56. G.B. Sweet and M.P. Bollman 1971: Variation in seed yields per cone in Douglas fir in New Zealand.**
This interesting study revealed that number of full seeds per cone was much lower from most NI sites than SI sites, except Ngaumu. Only a low proportion of ovules were fertile in Rotorua and Hawkes Bay.

**59. C.J.A. Shelbourne, J.M. Harris, J.R. Tustin and I.D. Whiteside 1973: The relationship of timber stiffness to branch and stem morphology and wood properties in plantation-grown Douglas-fir in New Zealand.**
Thirty-two sample trees were selected from a 45-year-old unevenly stocked, untreated stand of Douglas-fir in Kaingaroa Forest. They were selected in eight character groups, with two trees for high and two for low values of each trait. These groups were stem kinks, stem sweep, branch diameter, branch angle, number of branches per cluster, number of clusters, wood density and stem diameter.

Four 16 ft. logs were sawn from each tree to produce 2 in. material which was visually and machine stress graded. All possible morphological characters were measured on each tree and wood properties measured on discs at top and bottom of each log.

This study was directed at identifying the most important traits which controlled timber stiffness, the main trait of interest. Density variation was restricted by the single stand sampled. Wide differences in other traits were shown. Branch cluster frequency and branch number were more consistent in each tree than nodal swelling and stem deviations.

Average MOE of all pieces from each log was highly variable between trees, both in large dimension 8 × 2 in. and 12 × 2 in., and after resawing to 4 × 2 in., but variation between logs within trees was small. Trees with high MOE have small branches, small stem diameters, straight stems and high wood density.

Stepwise multiple regression of log average MOE showed that 46–71% of variation in stiffness (MOE) was accounted for by branch diameter. MOE was also accounted for strongly by nodal swelling, which was measured directly on each log as difference between diameter over and beside the biggest nodal swelling of each log. Wood density and stem deviations also accounted for MOE. This meant that quality of timber of a tree could be quite precisely predicted by these three traits. From this information selection of trees for breeding could be planned, with some added knowledge of the traits' heritability.

***Comment*** At the time, no consideration was given to stiffness (MOE) being a trait in its own right and related to micellar angle, as became apparent much later.

**55/6960 M.D. Wilcox 1974: Stress grading study of Douglas-fir; effects of branch diameter and wood density on timber stiffness.**

This study was set up by the silviculture group, including sampling. Stiffness of green 4 × 2 in. timber as a plank was strongly influenced by branch diameter and wood density. Branch diameter effects were 2.5 times greater than density effects on stiffness. The effects of branch diameter and density were linear and additive and account for 80% of variation in stress grade. Branch diameter alone accounted for 66% of that variation. Second log MOEs ranged from 0.925 psi × 106 for small branched logs to 0.584 × 106. Of total variance in stiffness, 50% was due to between trees and 50% within trees.

**69. M.D. Wilcox 1974: Douglas fir provenance variation and selection in New Zealand.**

*Comment* This was wrongly dated as 9/64. FTI report 18 by Sweet dated 1/64 gives nursery heights and flushing behaviour at 1 and 2 years from sowing of the Larsen 1959 seed collection in FRI nursery in January 1962.

Seed was collected in Washington, Oregon and California coastal sites by Egon Larsen in 1956. Trials were planted in 1959 at 19 sites (Sweet FTI Report 21) mainly 1 plot per provenance per site. For this assessment in 1973, 9 traits were measured on 15 of the 44 seedlots at 6 sites and on 45 seedlots, at 3 of those. Fifteen provenances were chosen on the basis of Thulin's statement in the 1966 NZFRI Annual Report, possibly on nursery heights. There is no report of a previous field assessment of the 1959 trials, though Wilcox gives 1965 heights (age 6) based on SI and NI trials' assessments in a Douglas fir provenance variation and selection in New Zealand.

The research sampled gives sampling of trees per provenance-location cell as 42 for DBH, height, straightness and malformation with 15 of those scored for needle retention, branch diameter, wood density and heartwood %. Survival was based on counts of the 64–192 trees per plot planted (see Sweet Report 21 for number of plots and number of trees per plot).

Analysis was based on a two-way crossed classification (provenances and locations) random model, with two levels of nested sampling, with unequal numbers of plots per location-provenance cell. Expected MS (mean squares) are provenances, locations, Prov. × Loc., plots in cells and trees in plots. Variance components were estimable for Provs., Locs., Prov. × Loc.

For the 45-provenance analysis, the provenance component for DBH was 7% of total variance and provenance × location component was 5%. This indicates quite high GxE. For height this was 27% versus 4% for provenance × location, much lower. For the 15-provenance analysis, DBH was only 1% of total variance and provenance × location was 1%.

***Comment*** I believe the unbalanced design with every variable (mostly one to three plots per provenance per site) and choice of provenances contributed to these results, reducing precision of estimation of provenance effects at each site. Fifteen trees per plot were assessed for some of the traits and 42 for others, DBH, height, straightness and malformation.

Repeatabilities of provenance means for 45 provenances were 0.90 (0.81) for height, 0.74 (0.56) for DBH, 0.75 (0.82) for straightness, 0.23 (0.71) for needle retention, 0 (0.14) for branch diameter, 0.70 (0.50) for heartwood % and 0.59 (0.82) for density (the 15-provenance analysis are in parentheses).

***Comment*** There was no attempt made to relate the earlier nursery height and flushing to these results which was unfortunate when one examines the nursery heights, which varied widely and came from a replicated nursery design.

Selection index ranking of provenances generally followed growth rate. For the 45-provenance analysis, the index included height, DBH, straightness and density. Best provenances were 642 from Berteleda; 647 from Mad River, Korbel; 659 from Stinson Beach; 660 from Santa Cruz; 654 from Caspar (Fort Bragg); 641 from Fourmile, Oregon (Bandon); and 636 from Deadwood, Oregon. Rankings at Rankleburn furthest south were much the same, but the Cartwheel Hut site included Darrington, Washington; Castle Rock, Washington; 643 Jewel Springs; and 651 Eel River, California.

***Comment*** The comprehensive analysis of these trials and the important results could have impacted strongly on the breeding programme, which was well started in 1974. This assessment was carried out in 1972–1973, and DBH and height apparently correlated well with earlier nursery heights. The report text is rather summary in nature and doesn't do justice to the 13- to 14-year data (which were strongly predictive of later assessments, as it turned out).

The implications of the superior growth of the coastal Oregon and Californian provenances were not examined and applied to the breeding programme and then based on the probable Washington origin of the Kaingaroa population used for selection. The possibility of doing immediate selection in the provenance trials, as was done very much later, was not considered, probably because of conditioning in thinking that plus-tree selection needed to be very intensive and of mature trees (even though the '268' experience with radiata pine changed that). The importance of provenance variation for Douglas-fir breeding was not recognised then.

In retrospect, the provenance trial results were given less weight in favour of plus-tree selection and OP testing of progenies, both for *P. contorta* and Douglas-fir. Not enough careful thought was given about the choice of base population. The wrong decision was made to use the Kaingaroa population (Washington provenance) as the sole basis of the breeding programme, and this held up the Douglas-fir breeding programme for over two decades.

For the Douglas-fir provenance trials and initiation of breeding programme, provenance trials had been planted in 1957 and 1959 (see the summaries above). The 1957 trials were not too helpful as they did not sample the coastal locations or locations south of Snoqualmie, Oregon. However, the provenances of the 1959 trials were collected by Egon Larsen and were generally targeted at coastal locations. By 1969 these trials were 12 and 10 years from planting, and it was clear from Geoff Sweet's write-ups at several stages that the more southern coastal provenances were doing very well relative to provenances like Darrington from coastal Washington.

The improvement programme of this species proceeded in much the same way as that of radiata pine in the early 1950s but in the case of radiata selection was in a second- or later-generation hybrid mix of Ano Nuevo and Monterey populations. However, most Douglas-fir of the Kaingaroa stands were apparently originally from Washington, and the performance of (second-generation) Kaingaroa seedlots in the 1959 provenance trials was considerably inferior to that of (first-generation) coastal provenances from California and Oregon.

Unfortunately, Sweet's results (in his absence) were ignored or deemed unreliable even though at age 2 years in the nursery, the Kaingaroa seedlot was well behind coastal Californian seedlots in growth rate. At the assessment and 1974 write-up of the 1957 and 1959 trials, the superiority of the southern coastal provenances was obvious. But, of course by then it was too late.

An interesting aspect of Douglas-fir breeding was the realisation that wood density was likely to be an important selection criterion, as this species was grown primarily for structural timber. The importance of micellar angle and its relationship to stiffness was not known for some decades. This was given some scientific backing in the stress grading studies of late 1972 and 1974. Plus-tree selection however was in advance of these studies. In cruising the Douglas-fir stands in Kaingaroa, up to 50 years old, four or five well-formed trees in a stand were initially selected and cored, densities recorded, and the highest density tree selected. However, as Gerry Vincent and Derek Birt found, neighbouring compartments sometimes had very different densities, indicating that probably density could vary a lot among provenances. This became apparent in the 1959 trials much later (Fig. 5.1).

**M.J.F. Lausberg, D.J. Cown, D.L. McConchie and J.H. Skipwith 1995: Variation in some wood properties of *Pseudotsuga menziesii* provenances grown in New Zealand. NZJFS (25).**

Significant areas of *Pseudotsuga menziesii var. menziesii* (Mirb.) Franco (Douglas-fir) have been planted in New Zealand using a very small number of provenances. The resource is regarded as suitable for a range of structural products, with little information available on provenances in terms of wood quality. Two provenance trials were sampled: 34-year-old trees of 39 provenances across 6 sites (1959 series)

**Fig. 5.1** A plot of 144 trees of the 1959 provenance trial of Californian origin at Rotoehu

and 19-year-old trees of 6 mainly NZ seedlots across 3 sites. Wood density patterns were also examined for 10 provenances on 4 sites with 30 trees per provenance.

There was a strong site effect on properties measured, and correlations between growth rate and density were negative on all sites and at most five-ring group positions. Juvenile wood extended to between 10 and 15 rings from the pith. In the 34-year-old trial, significant differences were found both within and between sites for density, stem diameter and amount of heartwood.

Differences between provenances in resin contents were also significant. Two Californian provenances, Stinson Beach and Mad River, were found to be of consistently high density and good diameter growth on all sites. In the 19-year-old trial, there were significant differences within and between sites for density, but diameter differences were significant only between sites. The results indicate that the Ashley Forest provenance is superior, having good diameter growth and higher wood density.

**H. McConnon, R.L. Knowles and L.W. Hansen 2004: Provenance affects bark thickness in Douglas fir. NZJFS 34(1): 77–86.**
Some coastal Californian provenances of Douglas-fir have bark that is visibly thicker and more deeply furrowed than more northern and inland provenances. From a literature study, it was evident that these variations in bark thickness most likely constitute adaptation to spatial and temporal patterns of wildfires within the natural range of Douglas-fir. Six provenances from the latitudinal range of Douglas-fir in the Pacific Northwest of the USA (37°–48°N) (in NZ provenance trials) were sampled for bark thickness and compared with a New Zealand landrace (Kaingaroa) seedlot at two New Zealand trial sites (38° and 46°S). The analyses showed that Californian provenances had significantly thicker bark than both the Kaingaroa

(ex Washington) control seedlot and the Oregon and Washington provenances. The most southern provenance (Santa Cruz, California) had the thickest bark. Thus, there was a steady reduction in bark thickness with increasing latitude of the seed sources. The bark thickness of the Kaingaroa seedlot was not significantly different from the Washington and Oregon provenances.

The provenance variations in bark thickness caused a bias in under-bark volume estimates from volume function '*T136*'. Errors in volume estimation were greatest for Santa Cruz (+7.1%), Jackson State Forest (+2.8%) and Mad River (+2.0%). It is recommended that volume equation '*T136*' should be revised to account for differences in bark thickness with provenance. *This is very interesting and adds to the results of the 1959 provenance trials.*

***Comment*** The first three unpublished internal reports about plus-tree selection for density in the 1930s stands of Douglas-fir in Kaingaroa Forest are interesting and revealing. There is little in the literature about the selection and open-pollinated progeny testing programme. Wood density was included as a selection criterion by taking cores in 3–4 times the number of candidate, well-formed and well-grown trees and then selecting those with high density. The reports cover this operation. The second report looks at the big difference in density between neighbouring stands, planted 1 year apart but of different seedlots. Big gains could potentially be made apparently by provenance selection. The other references include one by Irene Steele that reveals the size of the plus-tree selection programme and seed collection in Douglas-fir. This is the first recorded account of the initial plus-tree selection in Douglas-fir. Plus trees were selected at Kaingaroa and Whaka forests in 1968. Seed of 125 progenies was collected in 1969 and sown in 1970 at FRI nursery. Bulked seedlots of three controls with 7–29 trees each and a Santa Cruz provenance lot were included. Heights of seedlings were measured at age 2 years. This is a good report of early growth but limited by the absence of replication in the nursery. They were extremely fortunate at collecting in such a good seeding year.

The other reports are mainly those on Douglas-fir for which no PDFs were available.

**103/7010 T.G. Vincent and M.D. Wilcox 1976: Height and health of seven-year-old Mexican, Californian and Kaingaroa origin Douglas-fir at Kaingaroa and Gwavas forests.**

At age 7, the Californian seedlot (ex Santa Cruz) was tallest, followed by *Ps. flahaulti* from Mexico, then *Ps. menziesii* from Kaingaroa and *Ps. macrolepis* from Mexico. A heavy infection by *Phaeocryptopus gaumannii* showed *Ps. flahaulti* best for needle retention. *Ps. macrolepis* was losing needles, and the Kaingaroa seedlot retained more needles than the Santa Cruz seedlot (note that the Kaingaroa lot is probably Washington provenance).

**Restarting the Douglas-fir breeding programme in 1988. (Shelbourne, Low, Gea and Knowles 2005).**

In spite of the lack of information from the OP tests of 1972, it became clear that the 1959 trials were a good selection resource for the coastal provenances. An assessment of the 1959 provenance trials in 1988 at age 29 showed that several of the Californian and southern Oregon provenances were superior to the Kaingaroa seedlot planted as a control.

Getting provenance means over sites from the 19-site field design is possible using least squares analysis. This provides the basis for estimating provenance means across sites to enable provenance selection. This indicated some sites with 3 plots per provenance as suitable for selection, bearing in mind that seed collection was from about 12 to 15 trees per provenance by the late Egon Larsen. A total of 186 trees were therefore selected from about 8 provenances.

**C.B. Low and L.D. Gea 1997: Estimation of genetic parameters for growth and form traits in Douglas-fir progeny test on four sites aged 23 years.**
Charlie Low and Luis Gea estimated genetic parameters for growth and form traits in 1972 OP progeny tests of Kaingaroa selections of Douglas-fir on four sites aged 23 years. This provided the first genetic parameters for mature Douglas-fir from the 1972 OP families and allowed selection of the best trees in the better provenances.

After selection in the 1959 provenance trials, at last the NZ Douglas-fir breeding population (as now it might be called) contained some of the best selections from the local Kaingaroa (Washington), Californian and southern Oregon populations. The 186 clones in the Waikuku archive of Proseed are being used as a breeding population (Superline A) through collection of OP seed from them. Superline B will be created by collection of OP seed from the trials of the 225 progenies in 11 coastal provenances collected by Charlie Low and Mark Miller in Oregon and California (Fig. 5.2).

**Fig. 5.2** The late John Miller examining young grafts of Douglas-fir at Proseed's Waikuku site

**C.J.A. Shelbourne, C.B. Low, L.D. Gea, and R.L. Knowles 2007: Achievements in forest tree genetic improvement in Australia and New Zealand: 5: Genetic improvement of Douglas-fir in New Zealand. Aus. For. 70, 1 pp. 28–32.**

Douglas-fir is New Zealand's second most important plantation species (104,000 ha. planted) and has been grown here since the late 1800s. The main establishment periods have been 1900–1935, 1950–1970 and 1990 onwards. Provenance trials of 35 and 45 provenances were planted in 1957 and 1959 with large plots of about 144 trees at 8 and 19 sites, respectively, but these had very limited or no replication of provenance plots. Seed collections were 'commercial' for the 1957 trials (and only as far south as Snoqualmie in Oregon) but, for the 1959 trials, were made by NZ breeder Egon Larsen, with a concentration on coastal fogbelt sites in Oregon and California.

Results at age 6 and 12 years from the 1959 trials showed clearly that the coastal Washington seedlots were growing much slower than coastal southern Oregon and Californian provenances. Recent studies of 22 coastal fogbelt provenance trials have confirmed these results. By age 14 years (1973), the 1959 provenance trials were showing clearly that the native provenances from the fogbelt of coastal California and southern Oregon were growing faster in volume than the NZ landrace (ex Washington) populations from which the Kaingaroa progenies derived. For this and other reasons, the breeding programme was put on hold for the next 14 years.

In the late 1980s, there was a rapid rise in prices of Douglas-fir sawlogs and a renewed interest by the industry in growing this species. In 1988 it was decided to reactivate the breeding programme, and 186 selections (age 29 years) were made in the best seedlots of the 1959 provenance trials, mostly in the coastal fogbelt plots of Californian and southern Oregon origin. Selections in NZ seed stands of Ft. Bragg, California, southern Oregon and Washington origin were also included.

The seed company Proseed NZ funded the operation, grafted the scions of the 186 selections and subsequently planted the grafts in their Waikuku archive. These selections, called 'Superline A', probably had a rather restricted genetic base as they came from provenance seedlots each based on about 12 trees, collected by Egon Larsen.

Later in 1993 an FRI expedition of Charlie Low and Mark Miller collected 240 OP seedlots from 21 provenances in coastal California and southern Oregon, constituting 'Superline B'. These were planted at one NI and two SI sites in 1996. The original 1995 plan was to have 15 sublines of 27 OP (open-pollinated) families each, with mixture of provenances in each subline. Selections from each subline would be polycrossed, to estimate the breeding values of each, and pairs of individuals would be mated to provide for forward selections for the future breeding population.

This pollination proved unworkable and the breeding strategy has been revised to be a completely open-pollinated one, both for GCA estimation and for AS (advanced-generation selection). With Superline A selections all grafted and mostly flowering at Waikuku, collection of OP seed there and planting tests of OP progenies will provide a breeding value (GCA) for each female parent and will produce seed (the same OP seed) for within-family selection for advancing the breeding population.

(By 2016 the Superline B trials will be aged 13 years from planting and nearing a suitable age for selection and grafting.)

Based on the two sawing studies of the 1970s and recent work on variation in stiffness of timber in a stand of Ft. Bragg, Cal. provenance, it is clear that mean tree stiffness of timber is very variable in a stand or provenance. Work on genetic variation in sonic velocity in the 1969 OP progeny test of Kaingaroa selections has shown that sonic velocity is moderately heritable.

The Ft. Bragg provenance has been shown to have exceptionally fast growth and will form a small elite breeding population which can be open-pollinated and subjected to intense within-family selection. The assessment at age 29 years in the 1959 planting of the Californian fogbelt and southern Oregon provenances has shown an 18% superiority in DBH over the Kaingaroa (Washington) landrace. Gains predicted from a deterministic simulation study (Shelbourne et al. 2007) for balanced within-family selection, comparing seedling and cloned families in the main breeding population, showed that gains from pollen mix crosses were highest across most heritabilities but that gains from OP families were only slightly lower.

Gains were always higher for OP than full-sib families. Gains for clones were always higher than for seedlings. Gains predicted from cloned forward selected seed orchards involved within and among family selection.

**C.B. Low, N.J. Ledgard and C.J.A. Shelbourne 2012: Early growth and form of coastal provenances and progenies of Douglas-fir at three sites in New Zealand. NZJFS 42:143–160.**

Douglas-fir is the main alternative species to radiata pine and is planted in modest areas, especially in the SI. Originally seed was obtained from Washington, and trials planted in 1959 demonstrated that the fastest-growing provenances were from coastal California and Oregon. Selections were made from these provenance trials in 1988, and later in 1993 seed of 222 open-pollinated families was collected by Low and Miller (1993) in California and Oregon from 19 coastal and 1 inland provenance.

These seedlots were planted in 1996 along with eight New Zealand landrace seedlots as controls. The latitude of the seedlots ranged from 35° 07′N in California to 44° 10′N in Oregon. The trial layout was a sets in replications design of 7 sets of ca. 30 progenies, each with 30 replicates of single-tree plots, and was planted on 3 sites, 1 in the central North Island, 1 at the northern end of the South Island and 1 at the southern end. Average tree height, assessed at age 4 years from planting, was 2.5 m in the central North Island and 2.2 m at both South Island sites, with the tallest trees reaching 4.9 m in the central North Island.

The tallest provenances came from between latitudes 38° and 39°N, especially the Fort Ross and Navarro River provenances in California. The New Zealand control seedlots all grew well. The NZ seedlots originally from Fort Bragg, California, were the most vigorous. Seedlots that grew fastest were three inter-provenance crosses, planted in Southland only.

Height, stem straightness and needle retention at age 4 years showed moderate narrow-sense heritabilities (0.2–0.35). Tree form in these trials was highly variable with most of the variation in growth at the provenance level and somewhat less for families within provenances. Interaction variance of families × sites and also provenance × sites for height was moderate, but there were stable families that ranked well across all sites. Family means showed moderate correlation ($r = 0.6$) between height growth in the nursery and height at each site at age 4.

The trees in these trials form a very valuable resource for future breeding work. The first NZ breeding population of Douglas-fir was solely composed of Kaingaroa selections, probably Washington provenance, which were of limited use. From the provenances in the 1959 provenance trials, a relatively small number of trees were selected from the best provenances, as Egon Larsen only collected seed from 10–15 trees per provenance.

The seed collected in 1993 of over 200 progenies by Charlie Low and Mark Miller is an invaluable addition to the future Douglas-fir breeding population.

**C.B. Low, C.J.A. Shelbourne and D.G. Henley 2012: Effect of seed source of Douglas-fir at high-elevation New Zealand sites: performance at age eight years. NZJFS 42: 161–176.**

The trial involved the same seedlots as in Low et al. (2012) above from 11 native populations in coastal California, 3 Californian seed orchards, 1 Oregon seed orchard and 2 Washington seed orchards. Open-pollinated progeny seedlots were bulked by native population, and seed from seven New Zealand seed stands of Washington, Oregon and Californian origin was also included. This trial was conducted at five sites with a range of climatic features.

At the four trial sites with harsh climates, little difference was observed in the growth performance of trees from individual Californian provenances of varying latitude. Also, the three New Zealand seedlots of Californian origin grew at about the same rate as the three seed orchard seedlots of southern Oregon and Washington origin. At the four harsh climate sites, the severe effect of environment on stem straightness, malformation and acceptability appear to have largely obliterated any provenance differences at that assessment age (8–9 years). At the sheltered site, trees from northern Californian seedlots clearly grew best, and trees from the New Zealand seedlots originating from California substantially outgrew trees from the New Zealand seedlots of Oregon and Washington origin. This result paralleled that at three progeny trial sites which all had mild climates and included many of the same seed sources. The southernmost provenance from Los Padres at latitude 35° 49′ behaved differently from all other provenances. It had slower growth but better form.

**N.J. Ledgard and M.C. Belton 1985: Exotic trees in the Canterbury high country. NZJFS 15.**

A survey of exotic trees in the Canterbury high country showed that less than 0.1% of the 1.8 million ha. region is occupied by exotic trees. The major species present were Corsican pine, *Pinus nigra ssp. laricio*, ponderosa pine, *P. ponderosa*, radiata pine, *P. radiata*, European larch, *Larix decidua* and Douglas-fir.

A strong rainfall gradient was the major determinant of growth and, on average, could account for over 75% of the variability in wood production. In the moist zone, growth rates were good, with basal areas of over 130 $m^2$/ha. and volumes of over 1500 $m^3$/ha. being attained by age 40–50 years.

Maximum net annual increment ranged from <10 to >30 $m^3$/ha, depending on moisture availability. Other site factors such as slope, aspect and exposure appeared to influence growth but made minor contributions to the statistical analysis. Malformation (excluding butt sweep in larch) was worst in radiata pine (43% of all stems measured), larch (32%), Douglas-fir (21%), Corsican pine (18%) and ponderosa pine (10%). Wood densities tended to be low, in line with the national trend of decreasing density with increasing latitude and altitude. European larch showed the greatest incidence of spread of self-sown seedlings (62% of all stands), followed by Corsican pine (42%), ponderosa pine (37%), Douglas-fir (36%) and radiata pine (25%). The incidence of forest pathogens was low. Forestry is an efficient form of land use in parts of the Canterbury high country and has a definite role in any diversification away from traditional pastoral land use.

***Comment*** This was an interesting study showing relative adaptation of these important exotic species to rainfall and other factors. In the absence of *Dothistroma*, high-volume production is realised from most species in the moist zone. The good form of Corsican and ponderosa pines maybe means there is a future for Corsican, if ever these types of sites are needed for afforestation. There is a possible role for the *P. radiata* × *P. attenuata* hybrid.

## References


14. G.B. Sweet 1963: Some provenance differences in *Pseudotsuga menziesii*. 1. Seed characteristics.

21. G.B. Sweet 1964a: The establishment of provenance trials in Douglas fir (*Pseudotsuga menziesii*) in New Zealand.

23. G.B. Sweet 1964b: The assessment six years after planting of a provenance trial in Douglas fir (1957 series).

G.B. Sweet 1965: Provenance differences in Pacific coast Douglas fir. Silvae Genetica 14(2): 46–56.

37. M.D. Wilcox 1968: The genetic improvement of Douglas fir in New Zealand

56. G.B. Sweet and M.P. Bollman 1971: Variation in seed yields per cone in Douglas fir in New Zealand.

59. C.J.A. Shelbourne, J.M. Harris, J.R. Tustin and I.D. Whiteside 1973: The relationship of timber stiffness to branch and stem morphology and wood properties in plantation-grown Douglas-fir in New Zealand.

55/6960 M.D. Wilcox 1974: Stress grading study of Douglas-fir; effects of branch diameter and wood density on timber stiffness.

69. M.D. Wilcox 1974: Douglas fir provenance variation and selection in New Zealand.

M.J.F. Lausberg, D.J. Cown, D.L. McConchie and J.H. Skipwith 1995: Variation in some wood properties of *Pseudotsuga menziesii* provenances grown in New Zealand. NZJFS (25):

H. McConnon, R.L. Knowles and L.W. Hansen 2004: Provenance affects bark thickness in Douglas fir. NZJFS 34(1): 77–86.

103/7010 T.G. Vincent and M.D. Wilcox 1976: Height and health of seven-year-old Mexican, Californian and Kaingaroa origin Douglas-fir at Kaingaroa and Gwavas forests.

Restarting the Douglas-fir breeding programme in 1988. (See Aus.For. later: Shelbourne, Low, Gea and Knowles 2005).
C.B. Low and L.D. Gea 1997: Estimation of genetic parameters for growth and form traits in Douglas-fir progeny test on four sites aged 23 years.
C.J.A. Shelbourne, C.B. Low, L.D. Gea, and R.L. Knowles 2007: Achievements in forest tree genetic improvement in Australia and New Zealand: 5: Genetic improvement of Douglas-fir in New Zealand. Aus. For. 70, 1 pp. 28–32.
C.B. Low, N.J. Ledgard and C.J.A. Shelbourne 2012: Early growth and form of coastal provenances and progenies of Douglas-fir at three sites in New Zealand. NZJFS 42:143–160.
C.B. Low, C.J.A. Shelbourne and D.G. Henley 2012: Effect of seed source of Douglas-fir at high-elevation New Zealand sites: performance at age eight years. NZJFS 42: 161–176.
C.B. Low, C.J.A. Shelbourne and D.G. Henley 2012: Effect of seed source of Douglas-fir at high-elevation New Zealand sites: performance at age eight years. NZJFS 42: 161–1.
N.J. Ledgard and M.C. Belton 1985: Exotic trees in the Canterbury high country. NZJFS 15.

# Chapter 6
# Finale: Species and Provenance Testing of Conifers

A small 1955 trial of radiata pine provenances at Gwavas and Golden Downs forests involved two or three plots each of Ano Nuevo and Monterey populations of radiata, as well as OP progenies of '850'-7 and '850'-55. Otherwise the extensive provenance-progeny trials planted at two sites in Kaingaroa in 1964–1968 as well as later step-outs from these experiments at five other sites covered the genetics of this species well.

Douglas-fir (*Pseudotsuga menziesii*) is now the most important 'alternative' conifer, and provenance problems have been pervasive in the development of its breeding programme. As backup for the breeding programme, in 1993, Charlie Low and Mark Miller carried out selection and seed collection of over 200 OP seedlots in the coastal regions of California and southern Oregon. A clear provenance picture emerged in *P. nigra*. *Dothistroma* so depressed health and growth that there was no more planting. Provenance trials showed that ssp. *laricio* was top-ranked for height (and bole straightness) at all sites. However, the ssp. *laricio* Corsican provenances were highly susceptible to *Dothistroma pini* which put paid to this species for future use in the NI.

*P. ponderosa* var. *ponderosa* is subdivided into the Californian type and North Plateau type, with *P. ponderosa* var. *scopulorum,* and occupied 77,504 acres of all provenances and only 3000 acres of good Californian provenances. The impact of *Dothistroma* on all seedlots of *P. ponderosa* was disastrous, except in Otago and other dry, continental climates in the SI.

Provenance trials in *P. contorta* were planted in 1958–1961 at 16 sites and included ssp. *contorta*, ssp. *bolanderi*, ssp. *murrayana* and ssp. *latifolia*, and there were two NZ populations, Cpt. 51 Kaingaroa and Californian ssp. *bolanderi* from Waiotapu. The choice of the Waiotapu and Cpt 51 Kaingaroa strains as two separate breeding populations was probably wrong. A single breeding population, including all available coastal provenances from the trials, would have given more genetic gain at much lower cost.

Provenance trials were established intensively for *Larix decidua* (European larch) and *L. leptolepis* (Japanese larch).



C. J. A. Shelbourne, M. Carson, *Tree Breeding and Genetics in New Zealand*,
https://doi.org/10.1007/978-3-030-18460-5_6

*Pinus pinaster* (maritime pine) provenance seedlots were sown in 1953 and trials planted on nine sites. The species is widely distributed in Europe and a total of 47 seedlots were tested, from Portugal, Spain, France, Italy, Corsica and Morocco and two from NZ.

*Pinus muricata* (bishop pine), usually known as 'muricata' in NZ, was planted on a modest scale in Kaingaroa Forest in the 1920s–1930s where it was well adapted and gave some relatively high growth rates. Early on, it became clear that there were at least two chemical races, the alpha-pinene 'blue' and the delta-3 carene 'green', and the 'blue' race was evidently a distinct provenance, superior to the 'green' in form and growth. In the Bannister provenance-progeny trial at Rotoehu and other trials, the Sonoma County provenance (green) grew faster than the Mendocino County 'blue' population to the north and faster than the Marin County population to the south from which the 'green' stands in NZ had derived. The few 'blue' stands in Kaingaroa, planted in the 1920s–1930s period, looked magnificent, and heights were generally about 85% of radiata's. It was quite resistant to *Dothistroma*, but it will not hybridise with *P. radiata*, unlike its cousin *P. attenuata*.

*P. attenuata* was represented by a range-wide collection of seed with the provenances ranging from southern Oregon to south California and altitudes mostly 2500–4500 ft. There were two hybrid lots with *P. radiata*. The trials had to be destroyed because they were a focus of *Dothistroma* infection. The trial at Naseby, outside the *Dothistroma* area, was preserved, and the best trees there, about 15 of them, were selected and used as pollen parents in a trial of *P. attenuata* × *P. radiata* hybrids, which have shown superior growth, form and frost hardiness, on very cold sites.

The two Southern pines from SE, USA, *P. elliottii* var. *elliottii* (slash) and *P. taeda* (loblolly pine), tolerated the P-deficient gumland clays of Northland better than most species, but by 1970 they had been superseded by radiata in Northland. *P. taeda* at Rotoehu grew fast, but its form was ruined by possum damage and by low wood density and bark pockets beneath branches.

Provenance trials of *P. strobus* (77 provenance seedlots) were out-planted at three sites, Gwavas, Rotoehu and Golden Downs, utilising a 'sets in replications' design, with 20 provenances per set and 5 reps of 10-tree plots per site. Heights 5 years after planting showed that southern Appalachian provenances (from North Carolina, South Carolina, Georgia and Kentucky) grew the fastest, but its very low wood density and slow growth made this species unpromising in New Zealand.

# Chapter 7
# Experiment Design in Provenance and Progeny Trials

Planting exotics started in the late 1800s. Foresters from Europe were inclined towards the conifers used in the UK, like *Larix decidua*, *Pinus nigra* (Corsican pine) and *Pseudotsuga menziesii* (Douglas-fir). Sitka spruce was not included, though it later became the most important exotic in the UK. *Picea abies* was little used in NZ though a basic native species in Scandinavia and also Europe.

The species trials are listed by year of planting, the numbers of sites, plots/provenance/site, and trees per plot. A common feature of all trials from 1954 until about 1962 was that they mostly have a design that features very large plots of 100–300 trees, one to three plots/seedlot/site and usually a wide coverage in NZ of many sites. For *P. radiata,* in 1955, large-plot trials were planted at Gwavas and Golden Downs of provenances of *P. radiata,* from Monterey and Ano Nuevo, which had 2 or 3 plots per provenance and 100–200 trees per plot. Progeny trials of *P. radiata* in 1955 (29 OP progenies) and 1957 (about 20 CP progenies) both utilised row plots of 20 trees and 3 replications in a randomised complete block layout at 3 and 2 sites, respectively.

A problem for all these trials was the absence of a constant comparable control species. It would have been difficult to select such a species, as there was the possibility that it would suppress the target species or be suppressed itself. One solution might have been to plant a few large plots of a known provenance of Douglas-fir nearby. Another idea was to plant the 'best' provenances of several species together in comparative trials of several species.

There are 43 entries listed in the Appendix that covers all species planted and years of experiment. The emphasis is on the establishment of these trials, characterised by species, year of establishment, number of seedlots/provenances, number and identity of sites, number of plots/provenance and number of trees per plot.

A curious and frustrating aspect of the many trials that Ib designed from 1952 to 1962 is their use of very large provenance plots with up to 300 seedlings each and 1 to 3 replications per site (often only 1) of each provenance at each site. Even where he used up to three plots/seedlot/site for some of the provenances, these trials were unlikely to give adequate precision in ranking provenance means at that site because



C. J. A. Shelbourne, M. Carson, *Tree Breeding and Genetics in New Zealand*,
https://doi.org/10.1007/978-3-030-18460-5_7

of the imbalance in number of plots in different provenances. When data from these trials was analysed, the lack of replication, large plot size, extreme imbalance in number of plots per provenance at that site and the large amount of across-site variation meant that there were usually no significant differences detected between seedlots within sites. Only when the plot data were analysed across all sites or over regional groupings of provenances at a site did a pattern of provenance variation emerge, and then it was necessary to pool data across several sites to give a provenance mean. An inherent problem with trials with so few replications was that provenance x replication interaction (plot error) would often account for a large proportion of seedlot variance and make seedlot differences non-significant. Where several sites were involved, this same source of error could result in provenance x site interaction, again negating results.

So why did an experienced scientist design nearly all his trials this way? With 3 replications of 200-tree plots, this would require 600 plants per site per provenance. The number of sites per species was from 2 to 15, with an average of ca. 6. This was a colossal investment in planting stock, trial layout, planting and maintenance. The lack of or scanty replication is hard to explain. In fact, any breeder faced with these problems would economise on number of trees per plot and reduce the number of sites planted to ensure significant differences.

From a forest mensurationist's standpoint, large plots would 'last' till rotation age, despite thinning, if all silviculture were kept the same. Provided sites of a provenance were sufficiently numerous, a satisfactory result was possible. Ib's trials were rarely sufficiently numerous for this to be a rationale for his design.

There were some indications of changes from 1960 onwards in experiment design with Geoff Sweet's contribution. Martin Bannister's provenance-progeny trial, the 'Genetic Survey of *P. radiata*' with its single-tree plots in an interlocking block design, was a very advanced design deriving from Bill Libby of the Forestry Faculty at Berkeley, University of California, and the first single-tree plot trial in NZ.

Curiously, I had written a Proposition at NC State for my PhD in 1965 entitled 'Current misconceptions about field designs for progeny testing' in which very small plot size (even single-tree plots) and incomplete block designs were proposed for getting better precision in modern progeny tests.

A small progeny test of radiata pine was planted which included 16 crosses among 8 seed orchard parents as well as 16 OP seedlots collected from the same clones in the Kaingaroa orchard. These families were planted in a single-tree layout with an objective of estimating GCA (general combining ability) and SCA (specific combining ability) of several traits, and to see whether the performance of orchard OP progenies matched that of the CP progenies, and getting good estimates of general combining ability in a single-tree plot layout. The results confirmed the utility of GCA estimates from OP orchard families and of single-tree plots. Single-tree plots were subsequently adopted for most trials, especially progeny tests. The outside pollen contamination in the seed orchard progenies was very evident.

# Chapter 8
# Eucalypt (Hardwood) Species and Provenance Research and Breeding Programmes

**10. G.B. Sweet 1962: The future supply of high-quality eucalypt seed for New Zealand. An analysis of the problems involved.**
***Comment*** Sweet's document is a literature review and problem analysis about eucalypts, not only seed supply but the whole question of species and provenance selection and what traits to select for. Sixty references (of which 18 are by LD Pryor) were not a lot of help, as most were on vegetative propagation, flowering, seed setting and control crossing, nothing on progeny testing or breeding strategy. Harry Bunn's recommendations for farm woodlot planting (1961) were *E. muelleriana*, *E. pilularis*, *E. botryoides*, *E. saligna*, *E. scabra* (*globoidea or laevopinea?*) and *E. obliqua* for northern districts and *E. gigantea (E. delegatensis?)*, *E. regnans* and *E. fastigata* for southern districts (Sweet 1962).

**R. 40. I.J. Thulin and T. Faulds 1966: Grafting of eucalypts. NZ J. For. 8(4):664–667.**
Tip cleft grafting was developed for several eucalypt species, *E. saligna, E. botryoides, E. scabra (globoidea?)*, *E. microcorys*, *E. muelleriana*, *E. ficifolia* and *E. leucoxylon*. Seedling stocks were young and vigorous, 5–12 in. tall. Scions 1/10–1/8 in. diam. 6 in. below apex, tip cleft grafted onto stock stem of matching diameter (Thulin & Faulds 1966).

This indicates that early on, tree breeders were getting interested in the timber eucalypt species, such as stringybarks and *E. microcorys*.

Provenance testing was regarded as the first priority. Second was flower production of grafts and seed orchard management techniques and then wood properties. This was a first attempt at looking at the possibilities of eucalypt genetic improvement.

**R. 1379. M.D. Wilcox 1980: Genetic improvement of eucalypts in New Zealand. NZJFS 10(2): 3–359.**
Tree breeding, integrated with an active programme of species and provenance testing, was being conducted in *Eucalyptus botryoides*, *E. saligna*, *E. regnans*, *E. delegatensis*, *E. fastigata*, *E. obliqua* and *E. nitens*. The programmes involved



C. J. A. Shelbourne, M. Carson, *Tree Breeding and Genetics in New Zealand*,
https://doi.org/10.1007/978-3-030-18460-5_8

provenance trials and family tests to give a broad base of genetic variability of the different species and to form genetically improved local seed sources. Several other species were being tested on a small scale (note: *E. botryoides* and *E. delegatensis* later were not pursued).

Selection criteria varied somewhat among the seven main species, but health, fast growth, good stem form and branching characteristics and 'useful' wood were needed in all species. Genetic variation of practical significance has been found in the tolerance of *E. regnans* and *E. fastigata* to frost.

Considerable emphasis was placed on searching for strains of *E. nitens* which are less palatable to the *Eucalyptus* tortoise beetle, *Paropsis charybdis*. The breeding method used entailed selecting the best trees in the most promising provenances, followed by intermating among these to produce improved seed, via seed stands, seedling seed orchards, clonal seed orchards and progeny-tested seed trees.

***Comment*** This was an introduction to the strategy of planting native open-pollinated seedlots from a variety of provenances of each species that went on to form the basis of a breeding population. Usually some plus tree selection and open-pollinated (OP) seed collection was done in stands that were already growing in NZ, and these added to the breeding population (BP). The 'breeding method' is non-committal and doesn't say where the provenance populations come from or how intermating is carried out.

Of the seven species listed above, only *E. fastigata*, *E. nitens* (in the SI) and just possibly *E. regnans* are still in breeding programmes. Results of pulping research indicate *E. regnans*, *E. fastigata* and *E. nitens* are inferior for kraft pulping. This implies that eucalypt programmes should proceed slowly to allow variation in growth performance, health and wood quality to be properly evaluated. A criticism of the eucalypt programmes in general was that Mike Wilcox was apparently in a hurry to get publishable (and immediately applicable) selection results. There wasn't much serious debate amongst the other breeders about the eucalypt programmes and breeding strategy.

Twenty-four species are listed as under test in species trials: 7 ashes, karri from WA, 2 stringybarks, blackbutt and *E. pyrocarpa* and the rest gums. Few of these species were properly tested in species trials. The 24 species are:

*Gums: E. cypellocarpa* (round-leaved gum), *E. dunnii* (white gum), *E. globulus* ssp. *globulus* (Tasmanian blue gum), *E. grandis* (flooded gum), *E. gunnii* (cider gum), *E. johnstonii* (Tasmanian Yellow gum), *E. viminalis* (manna gum), *E. diversicolor* (Karri, WA), *E. jacksonii* (red tingle)

*Stringybarks and ashes: E. globoidea* (white stringybark), *E. muelleriana* (yellow stringybark), *E. microcorys* (tallow wood), *E. pilularis* (blackbutt), *E. pyrocarpa* (large-fruited blackbutt). *E. andrewsii* (two ssp. ash), *E. dendromorpha* (Budawang ash), *E. fraxinoides* (white ash), *E. oreades* (blue mountain ash), *E. sieberi* (silvertop ash), *E. stenostoma* (Jillaga ash), *E. triflora* (Pigeon house ash)

A bigger group of 60+ species was listed as under test by Hathaway and Bulloch in Hawkes Bay and Wairarapa, assessed at age 22 by Ruth McConnochie and I at Kahuiti near Masterton. Individual breeding programmes included:

*E. botryoides* clonal seed orchard of 17 selections planted at Woodhill and Rotorua in 1974/5. (No provenance or progeny testing and later abandoned)
*E. saligna* progeny and provenance tests of 43 native, 55 NZ families planted in 1976 (see Wilcox, Faulds and Vincent 1980 IUFRO Brazil) and clonal orchards planted in 1976
*E. regnans* (the biggest, see frosting reports)
*E. delegatensis* provenance trials planted in 1977 (38 native plus 52 NZ select families), later discontinued in face of competition with *E. nitens*)
*E. fastigata* (48 NZ, 73 native families and 6 South Africa families planted 1979) see Wilcox, Rook and Holden 1980 IUFRO Brazil for results of frosting study
*E. obliqua* (62 families from various provenances of this species were imported). Not clear whether and where this was planted
*E. nitens* (several provenance tests and 80 OP families of Victorian provenances) planted in 1979

***Comment*** This was an incredible effort over a short period from 1974 to 1979. Seven provenance-progeny trials with 100–250 families each of six species (*E. botryoides* excepted) were planted at several sites. The two 1980 papers for the Brazil meeting plus internal reports and project records should be consulted to get the full story. *E. nitens*, *E. regnans* and *E. fastigata trials* should be documented in Eucalypt Breeding Cooperative reports.

**R. 1295. M.D. Wilcox 1979: The ash group of eucalypts. NZJFS 9(2): 133–144.**

This was a general description of the ashes from the sub-genus *Monocalyptus* and series Obliquae. Best known ashes in NZ are *E. delegatensis*, *E. regnans*, *E. fastigata* and *E. obliqua*. These are tall forest trees of SE Australia and most important for sawn timber and paper making. They are pale coloured and low density compared with other eucalypts.

All described species are listed and discussed, and the more important species in NZ are given a bit more description. Obliquae includes ashes, mallee ashes and snow gums. Four peppermints, five scribbly gums and two stringybarks were provisionally added by Brooker, which classification Mike follows here (but not used later by me).

*E. obliqua* is the most widely distributed of timber ashes from northern NSW to south Tasmania and found in wet to dry sclerophyll forest. It has not grown consistently well in NZ. Provenances are under test in NZ. Good in East Coast sites.
*E. delegatensis* comes from high elevations in the southern highlands of NSW and Victoria, and from Tasmania, and was widely planted in NZ. It is hardy to frost and easy to establish.

*E. regnans* is the tallest of all plants and found below 1000 m in southern Victoria and below 600 m in Tasmania. Pulpwood plantations in Tokoroa have utilised S. Tasmanian provenances, but there have been past failures from frost. Genetically improved fast-grown seed sources are under development. *E. regnans* appears to be the best eucalypt for areas with cool sites and little frost.

*E. fastigata* grows best at 900–1200 m in NSW and is a major wet sclerophyll forest component in eastern NSW tablelands and escarpments. It has been widely planted in NZ, is the most adaptable and healthy of the ashes and is under a breeding programme to fix coarse branching.

Ashes: *E. oreades*, *E. lehmannii*, *E. pauciflora* (snow gum) (three ssp.), *E. gregsoniana*, *E. fraxinoides*, (white ash), *E. triflora*, *E. dendromorpha*, *E. obtusiflora*, etc. *up to no. 42.*

Better known varieties are *E. paliformis*, *E. sieberi*, *E. andrewsii* and *E. rossii*. (Wilcox 1979).

**R. 1298. M.D. Wilcox and I.J. Thulin 1979: Growth of *Eucalyptus regnans* in a plot at Rotorua. NZJFS 9 (2): 166–169.**

A nice plot is all that's left of a small provenance trial in the Long Mile, which did grow well. Seven altitudinal seedlots from Mt. Erica, Vic. from 366 to 991 m. Initial spacing was 1.53 × 1.22 m or 5357 stems/ha and thinned progressively to 237 stems/ha at 6 years.

At age 13 years, mean height was 23.7 m, mean DBH 41 cm and MAI 24 $m^3$/year (Wilcox & Thulin 1979).

**R. 1307. M.D. Wilcox, T. Faulds, T.G. Vincent and B.R. Poole *Eucalyptus regnans* F.Muell. Aust. For. Res. 10:169–84.**

One hundred forty-four open-pollinated families of 1-year-old *E. regnans* seedlings of several native provenances and NZ stands; in fact the breeding population were first tested for tolerance of frost in field trials at Wiltsdown (Tokoroa) and Kaingaroa. Frost damage varied widely among families, with 0–50% affected. Family variation also varied within provenances. There is probably a narrow genetic base in some NZ stands (Wilcox 1307).

The genetic improvement program was started in 1977 (see GTI Branch Report 75) with 144 families and 36 provenance bulked lots. Families are in four semi-isolated sets of 36, each representing possible future breeding populations and seedling orchards: Set A, Tasmania, Ellendale, Uxbridge, Maydena and Strathblane; Set B, Victoria, Narbethong, Mt. Erica and Strzlecki; Set C, NZ, selected parent trees from 13 places; and Set D, mixed origin, of Victoria, Tasmania and NZ, + extra four provenances.

The design was separate neighbouring locations of the four sets, A, B, C, and D, at Wiltsdown and Set D at Kaingaroa Cpt. 36, so the Tasmanian and Victorian provenances were kept separate. This was not a sets in reps design because the sets were chosen by populations. Spacing was 4 × 3 m, 36 reps of STPs. Frost damage was scored on 10 August 1978 from 0 (none), 1 (very slight) to 5 (dead). Analysis was a full-blown one with variance components, done one set at a time. Set C was NZ OP families of 'plus' trees, selected by Mike Wilcox and Gerry Vincent, of

**Fig. 8.1** *Eucalyptus regnans* breeding population of OP families at Wiltsdown, NZFP p. 11/77. Shown here as at 2/78

mostly unknown original provenance. Trees selected from Waitati near Dunedin were considered NZ's finest stand of eucalypts and performed the worst for frost (Fig. 8.1).

Percentage of trees severely frosted is variable (and not statistically comparable). Set A averaged 8.6% frosted (Tasmania), Set B 22.5% (Victoria), Set C 16.2% (NZ) and Set D 12.3% at Kaingaroa (mixed). Repeatabilities of family means were, respectively, 0.33, 0.61, 0.80 and 0.83. Frost score: mean scores by sets were, respectively, 1.11, 1.56, 1.28 and 0.88. Repeatabilities of family means were, respectively, 0.42, 0.70, 0.86 and 0.88.

Families were ranked within each set. Set A, Tasmanian provenances, ranged in % severely frosted from 0% to 20% and frost score varied from 0.60 to 1.68, mean 1.11; Set B, Victorian provenances, score ranged from 0.61 to 2.41, mean 1.56, worse than Set A with 1.11. Set C, NZ families, ranged from 0.39 to 2.51 mean 1.28, probably a mixed bag of provenances. Set D, the mixed lot, ranged from 0.13 to 2.45.

Mike Wilcox disagreed with the theory that these eucalypt families were half-sibs with one quarter of additive genetic variance among them. He noted that the design meant that differences in provenance were confounded with family differences, which themselves varied in inbreeding, in turn inflating family variance. Family means (unfortunately) did not provide a reliable measure of GCA of mothers. He recommended that in this first generation, emphasis be given to frost tolerance which is less likely to be disturbed by inbreeding effects than growth.

The immediate application of results was to collect seed from the best NZ mothers for commercial planting, with future seed collected from resulting stands. Mike realised however that provenance selection might be more efficient in improving frost resistance, but the Tasmanian and Victorian provenances could not be compared. Family test results will be used to thin tests and convert them to SSOs (seedling seed orchards). A clonal orchard had already been established (with NZ selections) and would need roguing. There was no decision for *E. regnans* about the composition of the breeding population regarding provenance.

***Comment*** This paper was published early in the *E. regnans* breeding programme and did open up a lot of the issues in the genetic improvement of eucalypts, such as provenance, family structure, native vs NZ populations and traits for selection. The development of breeding, testing and seed orchards was done in a hurry, and selection decisions, such as against frosting, were probably too early. Relative importance of species, traits, families and provenances all can change a lot with time.

**R. 1308. D.A. Rook, M.D. Wilcox, D.G. Holden and I.J. Warrington 1980: Provenance variation in frost tolerance of *Eucalyptus regnans* F. Muell. Aust. For. Res. 10:213–238.**

For this frosting study, 38 provenance seedlots of which 23 were from Griffin's range-wide collection of 49 provenances were supplied by Australia's CSIRO (Commonwealth Science and Industrial Research Organisation). These were based on from 7 to 49 parents in each case. Three lots were from APM (Australia Paper Mills) Traralgon, Victoria, of bulked plus trees; Ken Eldridge supplied a Mt. Erica provenance lot of trees from various altitudes on Mt Erica, Victoria; and there were two commercial collections from Tasmania and two from Victoria.

Seed was sown in December 1978 and seedlings were frosted in May 1979, July and October. At each season five or six runs/replications with 2–3 seedlings per provenance were run at three frost levels. An average of 45 seedlings per provenance were frosted in May, 48 in July and 31 in October. Frost damage was assessed on a 0–5 scale, as for the field trial.

The most frost-tolerant provenances were from interior, upland sites in south central Tasmania, e.g. Moogara, Styx river, and from high elevations (>900 m) in Victoria. Intermediate frost hardiness was shown by provenances from altitudes 500–800 m and by south central and south-eastern Tasmanian provenances. Southern Victoria provenances from the Otway ranges and Strzelecki ranges in Victoria, and northern Tasmania provenances were least frost hardy. NZ seedlots varied from very hardy to very tender. Provenance rankings were very similar to those in field trials.

Recent species trials indicate frost tolerance ranking is *E. nitens*, *E. delegatensis*, *E. fastigata*, *E. regnans* and *E. obliqua*, with pronounced provenance variation in each. Hardiest Moogara, Tasmanian *E. regnans*, is better than some provenances of *E. fastigata*, but the hardiest *E. fastigata* provenances are far more frost tolerant than the best Moogara provenances of *E. regnans*. These hardy *E. fastigata* from high elevation sites in NSW are hardier than some provenances of *E. delegatensis* from Tasmania.

**M.I. Menzies, D.G. Holden, D.A. Rook and A.K. Hardacre 1981: Seasonal frost-tolerance of *Eucalyptus saligna, E. regnans* and *E. fastigata.* NZJFS (11).**
Seasonal frost tolerance of *Eucalyptus saligna*, *Eucalyptus regnans* and *Eucalyptus fastigata* was determined by frosting seedlings monthly during 1976 in a controlled environment. Frost tolerance of *E. saligna* ranged from ca. −3 °C in summer to −7 °C in winter, of *E. regnans* from ca. −3.5 °C in summer to −9 °C in winter and of *E. fastigata* from ca. −4 °C in summer to −10 °C in winter (Menzies et al. 1981). Provenance of all these seedlots is not stated. Seedlings started hardening in April; *E. saligna* reached maximum tolerance in late June, whereas *E. fastigata* reached maximum frost tolerance in late July, the coldest part of the winter. *Eucalyptus saligna* started to deharden during July, whereas *E. fastigata* started to deharden a month later, and all three species continued to deharden throughout October.

The frost tolerance of the three eucalypt species is less than that of *Pinus radiata* (D. Don) for all times of the year, and so their establishment should be restricted to sites with good air drainage to minimise the risk of frost damage.

**M. Dick:1982 Leaf-inhabiting fungi of eucalypts in New Zealand. NZJFS (12).**
Of the leaf-spot diseases of eucalypts recorded in New Zealand, only those caused by *Mycosphaerella cryptica* (Cooke) Hansford, *M. nubilosa* (Cooke) Hansford and *Septoria pulcherrima* Gadgil and Dick are considered to be of any significance. Other fungi reported are *Aulographina eucalypti (*Cooke & Massee) von Arx and Muller, *Cercospora eucalypti* Cooke and Massee, *Hendersonia* spp., *Microthyrium eucalypti* P. Hennings, *Phaeoseptoria eucalypti* Hansford, *Trimmatostroma bifarium* Gadgil and Dick and *T. excentricum* Sutton and Ganapath (Dick 1982).

*Mycosphaerella* is one that caused serious problems for *E. regnans* and for *E. nitens.*

**R. 1503. M.D. Wilcox 1982: Anthocyanin polymorphism in seedlings of *Eucalyptus fastigata* Deane et Maid. Aust. For. Res. 30:501–9.**
This study is based on 105 families from 10 seed sources (only 4 native), Barrington Tops (7), Oberon (10), Robertson (11), Bombala (13), South Africa (6), Cambridge NZ (5), BOP NZ (17), Oakura NZ (11) and Hunterville (9).

Frequencies of purple seedlings varied by NZ population from 0.01 for Hunterville to 0.31 for Oakura and in native Australian populations from 0.19 for Robertson to 0.93 for Rossi. Frequency of green phenotypes varied: 0.02 for Rossi to 0.77 for Robertson and 0.1 for SA to 0.99 at Hunterville.

This demonstrated extreme genetic (provenance) variability. Family mean repeatability within populations was 0.98. Trials were planted in 1979. Population variance component was 50–55% versus families-in-populations of 10–14%.

**R. 1604. M.D. Wilcox, T. Faulds and T.G. Vincent 1980: Genetic improvement of *Eucalyptus saligna* Sm in New Zealand. IUFRO Symposium and Workshop, Aguas de Sao Pedro, Sao Paulo, Brazil Aug. 1980 (also published 1982 by FAO Rome in Forest Genetic Resources Information No 11).**
There were 18 provenances of *E. saligna* planted on four sites in 1976 and assessed in July 1979. A table gives mean DBH, height and stem quality of 14 provenances

plus one of *E. botryoides*, three of *E. grandis*, two of *E. deanei* and one of *E. dunnii* at age 3 years, averaged over three sites. Also included are reports of planting of clonal seed orchards of *E. saligna*; 78 NZ plus trees were selected in 1975/1976, 40 of these were grafted, and small orchards planted at Tairua and Kauaeranga.

***Comment*** With hindsight, this programme in which local trees were selected and grafted into orchards, and open-pollinated progeny tests established of these selections, was 'too much' 'too soon'. This species was still insecure in relation to other species and to utilisation and demand. When combined with parallel trials of native provenances and local bulked seedlots, the programme had evidently advanced too fast. Results were acted upon when the trials were very young and small (50–70 mm DBH.) Incidentally, *E. saligna* performed very poorly relative to *E. fastigata* and *E. regnans,* much later, in assessing eucalypt species trials in Northland.

**M.D. Wilcox 1982: Genetic variation in frost tolerance, early height growth rate and incidence of forking among and within provenances of *E. fastigata*. NZJFS (12).**

One hundred and twenty-six seedlots (115 open-pollinated families and 11 composites) of *Eucalyptus fastigata* Deane & Maid, representing 8 native provenances from New South Wales and Victoria, 1 exotic population from South Africa, and 15 exotic populations from New Zealand were planted in tests in 1979 at Kinleith and Kaingaroa. The trees were assessed at Kinleith in 1980 (age 1 year) for height growth and tolerance to winter frosts and at Kinleith and Kaingaroa in 1981 (age 2 years) for incidence of forking.

*Design*: 126 seedlots were divided into three sets of 42 seedlots with approximately equal representation of provenances in each set. There were 36 reps. of STPs at Kinleith and 42 at Kaingaroa. The design was 'sets in replications' (Schutz and Cockerham).

One-year height and frost damage were assessed in October 1980, 1 year after planting. Frost damage was on a scale of 0 (no damage)–5 (dead). Form pruning was done in February 1981 at age ca. 1.5 years, scored 100 if form-pruned or 0 if not. Provenance mean frost scores varied widely among provenances (including Oakura and Hunterville) from 0.02 for Barrington Tops to 0.11 for Robertson. Five Cambridge families showed very low frost scores of 0.03 and 17 Bay of Plenty families with 0.08. Oakura and Hunterville both showed heavy frosting with scores of 0.36 and 0.24.

Components of variance for 'provenances' were 3–4 times larger than components for 'families-in-provenance'. There were big provenance differences in forking and NZ seedlots and Robertson with 35–49% forking that were far worse than the other native provenances. The hardiest native provenances generally grew the slowest and showed the lowest incidence of forking. By far the hardiest native provenances were from Oberon and Barrington Tops, New South Wales. Families from New Zealand and from Robertson, New South Wales, were more frost-tender and more forked than those from other Australian localities and from South Africa. New Zealand families from Oakura and Hunterville showed excellent vigour but generally poor frost tolerance and a high frequency of forking.

The provenance in which the families possessed the best combination of good frost tolerance, fast growth and freedom from forking was from Bondi State Forest (south of Bombala, New South Wales) towards the southern end of the species' natural range. Future seed importations should concentrate on provenances from altitude 1000 m in the Great Dividing Range from Tallaganda southwards to the Errinundra plateau in Victoria. NZ populations showed little value for future use because of low frost tolerance and high forking, in spite of fast growth. There were quite high correlations between incidence of family mean forking and frost score, both within sites and between sites, ranging from 0.51 to 0.73. There were stronger correlations between provenance mean frosting and forking, 0.85–0.95.

***Comment*** Mike Wilcox did a great job of dissecting the data from this experiment. However, it was very early days in these trials for much to be learnt about *E. fastigata* breeding. It is a pity that this was nearly the last Wilcox report on eucalypt breeding.

**R. 1605. M.D. Wilcox, D.A. Rook and D.G. Holden 1980: Provenance variation in frost resistance of *Eucalyptus fastigata* Deane & Maid. IUFRO Symposium and Workshop, Aguas de Sao Pedro, Sao Paulo, Brazil. Aug. 1980. (also published 1982 by FAO Rome in Forest Genetic Resources Information No 11).**
This is a less detailed report on the *E. fastigata* frosting trials in growth rooms using the same procedure as for *E. regnans* (reported in Rook et al. 1980). This is a conference paper, yet provides a full tabulation of the mean frost scores and the % trees severely frosted.

Winter-timed frost gave best resolution. Native seedlots from the highest altitudes and possibly more interior and continental in climate had lowest frosting percentages. Barrington Tops, the northernmost provenance from 1370 m altitude, was 22.2% severely frosted; Oberon, from 1220 m altitude, was 29.4%, with the rest of the natives ranging from 60% to 85%. None of the NZ lots were lower than 67% though the South African lot from Draycott was 46%. The *E. regnans* controls were at best 85% severely frosted. Good agreement was found between *E. regnans* field frosting assessment and the growth room trials.

**R. 1590. M.D. Wilcox 1982c: Preliminary selection of suitable provenances of *Eucalyptus regnans* for New Zealand. NZJFS 12(3): 468–479.**
Thirty-six population seedlots of *E. regnans* were planted in 1980 on two sites, Wiltsdown, Tokoroa (altitude 260 m), and Kaingaroa (altitude 420 m) with eight NZ bulk lots from various sources and 28 Australian provenances, half from Victoria and half from Tasmania. These ranged in altitude from up to 1000 m at Toorongo down to 300 m at Narracan and 380 m at Wilsons Promontory, Victoria. In Tasmania, altitudes were lower, 150 m at Levendale to 600 m at Moogara (Wilcox 1982c). (These trials were additional to the family experiments at Wiltsdown and Kaingaroa, but the design and plot size have not been stated.)

Provenance mean heights at Wiltsdown varied from 7.6 m for Strzelecki, Victoria, to 5.9 m for Ferntree, Tasmania and 6.3 m for Yalmy R., Victoria. *Mycosphaerella* scores varied from 2.2 for a Tokoroa seed lot and from 2.6 for Mt. Erica, Victoria, to 4.3 for Rubicon, Victoria. Branching and straightness also varied with provenance and seedlot but not massively.

Kaingaroa heights were a bit lower, 5.2–6.5 m, with Strzelecki and Narracan provenances still top. There was a very small range of regional mean heights over the two sites, e.g. South Gippsland, 6.48 m, and high-altitude Victoria to 6.16 m. Regional mean heights of coastal east and southeast Tasmania provenances were 6.18 m and northern Tasmanian 6.33 m, NZ stands were 5.95 m (low), but two NZ seedlots were high, with Tokoroa ex Tasmania 6.45 m and Rangiwahia ex Victoria 6.85 m. There was a lot of variation among seedlots from Australia. Interior south Tasmania was marginally better for frost and disease resistance. NZ seedlots were also very variable.

In retrospect, there seemed little evidence at this early age to confine the breeding population to any particular provenance, and the local NZ stands were somewhat inferior as sources of selections. Frost resistance recorded in the first winter after planting showed highly significant variation among provenances. By age 3 years, there were significant provenance differences in height growth, resistance to *Mycosphaerella* leaf blotch disease, stem straightness and branching quality.

The most promising native Australian provenances were from southern Gippsland, Victoria (fast growth, good resistance to *Mycosphaerella*, but comparatively poor frost tolerance), and from interior southern Tasmania (satisfactory growth, reasonable resistance to *Mycosphaerella*, and excellent frost tolerance). A New Zealand seedlot collected from selected trees in a 9-year-old plantation of southern Tasmanian origin from Tokoroa also performed well.

There is a need to relate these results and those from the family tests to subsequent results. It is also interesting that *E. regnans* was virtually written off because of Barren Roads syndrome later. We need to check on results of the second-generation open-pollinated progeny tests which were badly affected by *Mycosphaerella*, like the one at Galatea. The status of *E. regnans* as a short-fibred kraft pulping species was downgraded by its low density and its low bulk at age 21 (see Kibblewhite, Riddell and Shelbourne 2000 in NZJFS30).

**R. 1591. M.D. Wilcox 1982d: Selection of genetically-superior *Eucalyptus regnans* using family tests. NZJFS 12(3): 480–493.**

The OP progeny tests of 141 families planted in 1977 were assessed at age 3 years from sowing in June 1980. Frost damage was scored in 1978 (Wilcox et al. 1980a, b). The field design was 36 reps of STPs (single tree plots) in four independent sets, (A) Tasmania, (B) Victoria and (C) NZ families, all planted at Wiltsdown, and (D) mixed origin, planted at Kaingaroa. Analysis was done one set at a time at Wiltsdown, and at Kaingaroa Set D, separately. Height, disease, branching score and stem straightness score were all assessed.

The ANOVA used least squares analysis, and variance components were estimated for families, replications and error. Tasmanian family mean heights varied from

7.4 m to 6.2 m and Victorian family means from 8.5 to 6.8 m, with fast-grown families coming from a variety of provenances. *Mycosphaerella* scores averaged 2.6 for Tasmanian and 4.1 for Victorian provenances. Frost scores were slightly lower for Tasmanian (Wilcox 1982d).

A 0.8 ha clonal seed orchard was planted at Cambridge in 1982 with 42 NZ clones from selected plus trees, with 17 forwards selections (Australian) from within the progeny test and 17 forwards selections from NZ families. Sixty per cent of the clones were of Tasmanian origin, either native or NZ. The family tests were thinned (1983) to remove inferior trees and used for seed collection. Extensive Seedling Seed Orchards (ESSOs) of bulked seed of 30 NZ and 7 Australian families were planted.

***Comment*** This thinning seemed early and would reduce numbers in the OP progeny tests for further assessment. However, this programme was targeted at providing commercial seed quickly, with some knowledge of provenance and some degree of selection at the provenance and family level. In my opinion, the eucalypt breeding programmes should have been taken slower and more attention given to longer-term species and provenance testing and with fewer species under breeding programmes.

**FRI Bulletin No. 95. M.D. Wilcox, J.T. Miller, I.M. Williams and D.W. Guild 1985: *Eucalyptus* species trials in Longwood forest, Southland. Bulletin No. 95, FRI, NZFS, PB. Rotorua, NZ.**

The species trial at Longwood, a low altitude (75–100 m) sheltered site in Southland latitude 46° 11′, was one of rather few species trials planted in the initial burst of activity on eucalypts in 1977 onwards. The design was a randomised complete block with main plots of 15, 30 or 45 trees and three replications. Provenances were single-tree plots (STP) within the main species plots, whose size depends on how many provenances were represented. For any given species, a plot could be quite small, with 3.5 × 3.5 m spacing and no buffer rows between species. The replication of provenances was thus minimal, with on average a maximum of three replications × nine trees per five provenances planted. There were five or more provenances each for nine species and one to three provenances each for eight species. The 17 species were:

*Eucalyptus regnans, E. delegatensis, E. fastigata, E. oreades, E. fraxinoides, E. obliqua, E. sieberi, E. obliqua x E. regnans* (ashes)

*E. nitens, E. saligna, E. viminalis, E. dalrympleana, E. johnstonii, E. gunnii, E. dunnii, E. cypellocarpa, E. globulus* ssp. *globulus* (gums) (Wilcox et al. 1985)

***Comment*** This design leans on the relatively new gospel of STPs. Ib's early experience with provenance trials was with very few replications of massive plots of 144–196 trees. My experience in Zambia was that species plots should have at least 15 trees per plot at final stocking. Why plant species trials that are only going to be good for 5 years or so without buffers or capacity for long-term volume measurement? We did realise that STPs were useless for genetic gain or species trials and that row plots were only good for short-term comparisons with a species like Radiata.

These trials were planted in November 1977 and assessed in October 1981 at age 4 years. (DBHs only were assessed at age 6 in 1983.) *E. regnans*, *E. fraxinoides* and *E. nitens* were tallest, with mean heights of 5.6–6.2 m and DBHs of 59, 65 and 77 mm, respectively. Other species heights ranged from 3.7 to 5.3 m (*E. viminalis*). Only *E. regnans*, *E. delegatensis*, *E. fraxinoides*, *E. nitens* and *E. globulus* were rated for prospects as 'good'.

Provenance trials of *E. regnans*, *E. delegatensis*, *E. nitens* (families only) and *E. globulus* were also planted with, respectively, 36, 36 and 42 families and 5 provenances each, which gave an additional view of those species.

Results were presented in tables of seedlot means from an assessment at age 4 years and in a paragraph of text for each species (in order of merit) based on the DBH measurement and field impressions in 1983.

The ranking was *E. regnans*, *E. fraxinoides* and *E. delegatensis* (a bit slower than the first two); *E. nitens* was next but downgraded by early severe attack on juvenile foliage by *Eriococcus* and sooty mould. Adult foliage was slightly attacked by *Paropsis*. The *E. regnans x E. obliqua* hybrid, *E. oreades*, *E. globulus*, *E. fastigata*, *E. sieberi and E. obliqua* all grew well with variable form. *E. globulus* did surprisingly well. The rest were not inspiring. *E. saligna* was obviously out of its climatic range, but this site seems to have been amazingly kind to most species, compared with other forestry sites in Southland. *E. fastigata* was disappointing in view of its good performance in the NI. *E. fraxinoides* was a winner here. *E. nitens* and *E. globulus* were regarded with suspicion. The first three ashes, *E. regnans, E. fraxinoides* and *E. delegatensis,* were recommended. All this is interesting in the light of subsequent assessment and rise of *E. nitens.*

***Comment*** End-product capabilities were never mentioned. Only a few years later, they became drivers in the search for good short-fibred pulping species, the only end product even now that has much commercial interest for eucalypts.

**J.N. King and M.D. Wilcox 1988: Family tests as a basis for the genetic improvement of *Eucalyptus nitens* in New Zealand. NZJFS 18.**

Vigorous and healthy growth (16 m heights) was shown by 8-year-old *Eucalyptus nitens* (Deane & Maiden) Maiden families in NZ trials. Central Victorian families were 10–15% greater in diameter growth and 35–50% better in tree form scores than seedlots from eastern Victoria (Errinundra) and southern New South Wales at age 8 years. There were no large or consistent differences among the three provenances of central Victoria – Macalister, Toorongo or Rubicon.

The large variation between populations within the Toorongo provenance, showing the Mt Erica population top-ranked for diameter growth and the Upper Thomson River population bottom-ranked, supports the theory of introgression from the Errinundra variety. Multiple trait index selection across sites was used to choose the best 20 central Victorian OP families. Selection of these families should give gains of 7.5% for diameter, which equates to a 19% volume gain, over unselected central Victorian families.

Predicted gains in form score are 8% for these selected families. Genetic variability in resistance to wind damage was demonstrated, and this characteristic was also used in the selection of half-sib families. Half-sib family selection can be utilised for seed production gains, and there are methods of advancing the population in open-pollinated families (King & Wilcox 1988).

***Comment*** These are the assessment results of the first *E. nitens* native population OP trials planted at Mt Tongariro and elsewhere. They included mainly Victorian native population seedlots plus a few *E. denticulata* from Errinundra and a few from S. NSW. The OP seed collected from selections in these trials formed the basis of the next generation of OP families. Problems of the anomalous variances in native versus plantation selections were not foreseen at this point.

**R.L. Hathaway and M. King 1986: Selection of *Eucalyptus* species for soil conservation planting in seasonally dry hill country. NZJFS 16.**

The performance of 56 *Eucalyptus* species (119 provenances) at 2 exposed, seasonally dry, hill country sites in the Wairarapa district was assessed at age 5 years. There were significant differences among species in height and diameter growth, *Eucalyptus* tortoise beetle (*Paropsis charybdis* Stal) and leaf roller caterpillar (*Strepsicrates macropetana* Meyrick) damage, wind damage, stem straightness, crown width, crown density and branch size. *Eucalyptus cordata* Labill., *E. fastigata* Deane et Maid., *E. fraxinoides* Deane et Maid., *E. obliqua* L'Herit, *E. pulchella* Desf. and *E. regnans* F. Muell. ranked highly for most traits at both sites and are considered to be the most suitable of those species species tested (Hathaway & King 1986).

***Comment*** This is the first assessment at age 5 of the Kahuiti and Pakaraka trials, later assessed by Bulloch, and later still at age 22 by Ruth McConnochie.

**R. 2053. M.D. Wilcox and R.L. Hathaway 1988: Use of Australian trees in New Zealand. Proc. AFDI International Forestry Conference for the Australian Bicentenary 1988, Albury, New South Wales, Australia Volume III.**

This was a rather flippant and entertaining talk about the eucalypts and acacias in NZ. It was lightweight yet practical and realistic from two enthusiasts who can still see the wood from the trees. It was sound but derogatory about the lack of success with eucalypts (except by NZFP and their pulpwood growing of *E. fastigata* and *E. regnans)*, all seen from 1988 when Mike Wilcox was 'driving' the division following a brief spell from 1982 to 1985 as research field leader of of GTI (Wilcox & Hathaway 1988).

**R. 2467. P.G. Cannon and C.J.A. Shelbourne 1991: The New Zealand eucalypt breeding programme. IUFRO Symposium on Intensive Forestry: The Role of Eucalypts P2.0201 Productivity of eucalypts. Durban, South Africa 2–6 September, 1991.**

This was a 1991 status report of the various eucalypt species tests, provenance trials and breeding programmes in NZ. It also introduced the new 1-year-old Eucalypt

Breeding Cooperative. The history of the Wilcox-reported trials (mostly in NZJFS) is outlined up to 1990 of the first generation from the mid-1970s (Wilcox 1980, 1982a, b). Field trials were initiated from 1973 to 1979 including 7 species trials, 11 provenance trials and 19 progeny tests on some of these species:

*E. andrewsii, E. campanulata, E. cypellocarpa, E. dalrympleana, E. delegatensis, E. nitens, E. obliqua, E. regnans, E. saligna, E. sieberi, E. stenostoma, E. triflora, E. viminalis* and provenance tests of *E. botryoides, E. fastigata, E. obliqua, E. saligna, E. regnans, E. delegatensis, E. nitens*. (This programme was never actioned in full.)

Breeding population progeny tests were initiated for *E. fastigata, E. nitens, E. regnans, E. saligna* and *E. delegatensis*.

(See M.D. Wilcox, B.R. Poole and Fry 1980: The establishment of eucalypt plantations on the central NI plateau. Combined Conf. Institute of Foresters of Australia and NZIF.)

*E. regnans* and *E. fastigata* are the best species for altitudes of 300–500 m in CNI. *E. nitens* is probably the best for higher elevations in the NI and in the SI (refer also Bunn, Franklin, Moberley and Revell). There were several problems with establishment and health, with a lower success rate than radiata. Weeds, frost, insects, fungi, rabbits and possums were all problem-causing. There were special problems with *E. nitens* and *E. globulus*. Research on eucalypts decreased heavily during the 1980s, though New Zealand Forest Products continued to plant *E. regnans* and *E. fastigata*. There were 15,000 ha in total.

The Eucalypt Breeding Cooperative started with *E. regnans, E. nitens* and *E. fastigata* breeding programmes and some work on control pollination techniques, hybrids, flowering and seed production and species trials.

A 300 family second-generation open-pollinated breeding population (OPBP) for *E. regnans* had single-tree plot (STP) trials, 30 reps, sets in reps design on two sites for planting late 1991 (best second-generation NZ and some first-generation native families). They were also planted as sublined forwards selection blocks.

Three hundred second-generation OP progenies of *E. nitens* were planted in 1990 as two single-tree plot and one sublined forwards selection planting, with two additional trials of 60 families on frost sites (Cannon & SHelbourne 1991).

***Comment*** These were mainly Victorian provenances though NSW families have outgrown Victorian, with heavy Paropsis attack on Victorian. This concentration on Victorian provenances was a mistake, in my view, and so was concentrating test planting on lower altitude NI sites.

*E. fastigata* proved to be the best bet for resistance to insects and fungal dieback, but form was not good, and more emphasis was needed on branching and forking. There is plenty of provenance variation as well as family variation.

***Comment*** A great future was seen for interspecies hybrids, but first-of-all they need good cutting propagation from coppice. Lots of possibilities were seen by Phil and me (wrongly) based on my South African experience with clonal forestry.

Also cloning was a possibility for genetic testing and forwards selection. Propagation was the main barrier. In retrospect, we were dreaming. Clonal forestry has only worked in eucalypts for a limited number of species like *E. grandis* and *E. urophylla* in which coppice shoots root readily.

Species trials at various altitudes, included *E. fraxinoides*, *E. oreades*, *E. fastigata*, *E. regnans*, *E. saligna*, *E. globulus*, *E. grandis*, also *Populus deltoides*, *Acacia dealbata*, *A. mearnsii* and *E. macarthurii*, *E. nitens* and *E. delegatensis*, *E. obliqua* at higher/colder, *even E. gunnii* and *E. dendromorpha* at the coldest sites.

The entire future emphasis was on pulp, chemical and mechanical, but the needed wood characteristics were not yet known for these products.

**R. 2483. P.G. Cannon and C.J.A. Shelbourne 1993: Forward selection plots in breeding programmes with insect-pollinated tree species. NZJFS 23(1): 3–9.**
This describes a means of doing within-family selection for the future breeding population for insect-pollinated species with effective pollination ranges of about 40 m. The forward selection plots (FSP) contain about 30+ OP families of one subline, with 30–50 trees/family in row plots of 6–10 trees. Precise family rankings from another site with an STP design would give best precision. The row plot FSP design can estimate family means and facilitate within-family selection of the best tree per row plot. Subline blocks need to be isolated from one another (Cannon & Shelbourne 1993).

**J.N. King, R.D. Burdon and M.D. Wilcox 1993: Provenance variation in New Zealand-grown *Eucalyptus delegatensis*. 1. Growth rates and form. NZJFS 23.**
*Eucalyptus delegatensis* provenance trials, at two sites in New Zealand, were assessed at age 8 years for diameter and stem straightness. Tasmanian provenances overall had slightly larger ($p < 0.1$) diameters than Australian mainland provenances but were significantly ($p < 0.0001$) poorer in straightness than mainland ones. The results for diameter growth at age 8 years were in contrast to those for earlier (age three) height. At age 3 on two New Zealand sites (and on four sites in south-eastern Australia) Victorian provenances had clearly excelled Tasmanian but may later be overtaken by Tasmanian provenances (King et al. 1993).

New Zealand seedlots (commercial lots and open-pollinated families) showed, on average, modest diameter but good form, in line with their predominantly New South Wales origins. The families varied strongly in both diameter and form.

*Note* big changes in age 8 years results from age 3. Tasmania is best for growth now and for checking (see next report). *E. delegatensis* became inferior to *E. nitens* in SI and to *E. fastigata* in NI. STPs should *not* be used for provenance trials.

**R. 2528. J.N. King, R.D. Burdon and G.D. Young 1993: Provenance variation in New Zealand-grown *Eucalyptus delegatensis*. 2: Internal checking and other wood properties. NZJFS 23(3): 314–323.**
An 8-year-old trial of a range-wide set of provenances from Tasmania, Victoria, NSW and ACT at Longwood, Southland, was assessed for wood checking and density. Checking was general and serious, but Tasmanian provenances had <half the frequency of checking of 'Mainland' provenances. Tasmanian provenances had less

heartwood and density was 9 kg/m$^3$ higher and DBH very slightly lower than Mainland. Tree form was not as good as Mainland.

Relative to *E. nitens*, growth later proved inferior and *E. delegatensis*, though initially an important cold tolerant species in Southland, was replaced by *E. nitens*.

**M.J.F. Lausberg, K.F. Gilchrist and J.H. Skipwith 1995: Wood properties of *Eucalyptus nitens* grown in New Zealand. NZJFS (25).**
Some evaluations of solid wood samples (including internal checking assessment, shrinkage, collapse and tension tests) were carried out on 15-year-old *Eucalyptus nitens* trees grown in Kaingaroa. The trees were grouped in three 'classes' for low, medium and high stand density. The low-density class had significantly lower earlywood density than the medium and high stand density classes. Significant differences were observed between all three classes in average ring density and latewood percentage. The amount of internal checking observed is deleterious to most solid wood products. Collapse was frequent and severe, but steaming effectively restored the collapse. The scanner used gave an accurate and reliable estimate of radial and tangential shrinkage combined. The tension test results were positively correlated with density and moisture content (Lausberg et al. 1995).

***Comment*** A rather strange study on the effects of stocking on wood properties in *E. nitens*

**L.D. Gea, R.M. McConnochie and N.M.G. Borralho 1997: Genetic parameters for growth and wood density traits in *Eucalyptus nitens* in New Zealand. NZJFS (27).**
Genetic and phenotypic parameters for diameter at breast height, bole straightness, branching, pilodyn penetration and basic wood density were estimated for two open-pollinated progeny tests at age 5 years, including first- and second-generation material of different provenances. Different values for coefficients of relationship (0.25 and 0.5) needed applying in the heritability calculations for each generation, which accounted for the expected difference in their pollination behaviour and the observed difference in their variance component estimates.

Strong evidence was found for substantial realised gains from open-pollinated progenies from seedling seed orchards of North Forest Products (NFP) and AMCOR in Australia and a New Zealand Forest Research Institute progeny trial. Differences in diameter growth between the best seed orchard seedlot and the best native race were 0.67 standard deviations. In terms of wood density, the New Zealand progenies were clearly superior to those from Australian seed orchards. Families from AMCOR and NFP were of low to very low density (Gea et al. 1997).

***Comment*** This was the 5-year assessment of the second generation of E. nitens OP progenies, and OP family-derived parameters were compared of plantation versus native population selections. Coefficients of relationship were doubled for native lots, indicating full-sib relationship and/or inbreeding. This was the first indication of something weird with native OP lots, on which previous selection in various species was based.

**R. 2659. R.P. Kibblewhite, M.J.C. Riddell and C.J.A. Shelbourne 1998: Kraft fibre and pulp qualities of 29 trees of New Zealand grown *Eucalyptus nitens*. Appita 51(2): 114–121.**

This was the first of the individual-tree pulping studies on eucalypts with 29 trees, of which 9 were also individually pulped mechanically in the pilot plant. The information will be used in planning and management of end-product-directed plantation forests and tree breeding programs.

There was wide variation in density (390–556 kg/m$^3$) in the 15-year-old trees from a progeny trial of Victorian provenance. Kraft pulp yield of 54–59% varied widely and also kraft pulp properties of fibre dimensions and handsheet properties. Handsheet bulk, a key handsheet quality, was predicted by wood density and fibre length ($R^2$ 0.67) (Kibblewhite et al. 1998).

***Comment*** This information for the tree breeder meant there were possibilities for modifying these traits by selection of trees for wood properties. The studies themselves explored the whole field of relationships and prediction of pulping and handsheet properties using multiple regression. They were classic and very expensive research that depended on the variability among individual trees that allowed models to be built.

**C.B. Low and C.J.A. Shelbourne 1999: Performance of *Eucalyptus globulus, E. maidenii, E. nitens*, and other eucalypts in Northland and Hawke's Bay at ages 7 and 11 years. NZJFS 29 (2): 274–288.**

Species and provenance trials of eucalypts, planted independently, at Knudsen road, near Kaikohe in Northland and at Clive in coastal Hawke's Bay were assessed at ages 7 and 11 years, respectively. At Kaikohe, 11 provenances of *Eucalyptus nitens* (Deane et Maiden) Maiden, 6 of *E. saligna* Smith, 3 of *E. grandis* Hill ex Maiden, 2 each of *E. globulus* Labill, and *E. maidenii* Labill, and 1 of *E. robusta* Smith were included. At Clive, 21 species were involved, including *E. globulus, E. maidenii* and *E. bicostata* Labill, and also provenances of *E. nitens* from central Victoria (Vic) and southern New South Wales (NSW).

At Kaikohe three replications of 49-tree square plots were planted of each seedlot at 2.0 × 1.9 m spacing, and at Clive, a short rotation fuelwood trial consisted of 8 replications of 18-tree plots at 2 × 1 m spacing. At a third site at Patoka, northwest of Napier, transect comparisons were made of *E. nitens* (Vic.), *E. maidenii* and *E. bicostata* at age 11 years.At Kaikohe (age 7 years), the best growth was achieved by *E. nitens* (Vic.) and *E. nitens* (NSW) (equal; DBH 200 mm), followed by *E. globulus* and *E. maidenii* (equal; DBH 172 mm). However, the *E. nitens* (Vic.) provenances showed a widespread, unidentified disorder in the form of progressive loss of the lower crown and eventual death of the tree. Their "survival" (after a 50% early thinning) averaged half that of the NSW provenances which had healthy crowns. *Eucalyptus globulus* showed a similar problem to *E. nitens* (Vic.), but *E. maidenii* retained good crowns and high survival.

At Clive (age 11 years), a periodically high watertable affected survival and health of some species more than others. *Eucalyptus nitens* (Vic) grew well (DBH

200 mm), though survival was poorer than that of the NSW provenance and crown health was much poorer, with the same progressive loss of lower crown and death (also windthrow) as was seen at Kaikohe. *Eucalyptus globulus* had grown a little better than both *E. nitens* provenances, but showed similar crown death to *E. nitens* (Vic.). *Eucalyptus maidenii* had a higher mean DBH (219 mm) than all other species and had maintained a high survival. At Patoka, growth, form and health of *E. nitens* (Vic) and *E. maidenii* were excellent, with *E. maidenii* a little slower-growing. The good health and much higher wood density of *E. maidenii* than of *E. nitens* indicate its potential value for pulpwood, e.g. at Clive (from a disc study at age 10 years), weighted whole-tree basic density of *E. maidenii* was 582 kg/m$^3$ as against 450 kg/m$^3$ for *E. nitens* (Low & Shelbourne 1999).

***Comment*** Some 15 years later, this seems a good account of the relative performance of these eucalypt species. *E. nitens* has really failed in coastal BOP and Northland and the better performance of the NSW provenances does not persist. *E. maidenii* was too dense for pulpwood unless on very short rotations (as seen in a later kraft pulping study).

**C.J.A. Shelbourne, S.O. Hong, R.M. McConnochie and B. Pierce 1999a, b: Early results from trials of interspecific hybrids of *Eucalyptus grandis* with *E. nitens* in New Zealand. NZJFS 29(2): 251–262.**

*Eucalyptus nitens,* a high-altitude species from the mountains of central Victoria and eastern New South Wales, is commonly planted for pulpwood in New Zealand. Parents of these two provenances of *E. nitens* were crossed with *E. grandis* of coastal New South Wales origin to create first-generation hybrids that should be adapted to warmer, low-altitude sites in New Zealand and which might be able to be propagated by cuttings. Single-pair crosses were made between eight New Zealand-selected parents from both central Victorian and eight from southern New South Wales provenances of *E. nitens*, as pollen parents, with eight selected *E. grandis* female parents growing in a South African seed orchard. Eight seedlings of each hybrid family and of open-pollinated families from each of the *E. nitens* and *E. grandis* parents were planted in a single-tree plot design at each of four sites in New Zealand (Shelbourne et al. 1999a).

One trial in Southland was destroyed by frost, and hybrids performed poorly at a frosty central North Island site. At two warm, coastal Bay of Plenty sites the hybrids at age 2 years and 8 months had average height and diameter that was about the same as their *E. nitens'* open-pollinated siblings. There were however a large proportion of poorly grown, genetically defective individuals in the hybrid families and a smaller proportion of extremely vigorous trees which exceeded the growth of the best-grown individuals of *E. nitens*. These successful hybrid genotypes could form the basis of clonal forestry deployment of the hybrids, provided vegetative propagation methods can be developed.

The *E. grandis* progenies had a mean DBH at Te Teko of 114 mm vs 143 & 138 mm for hybrid Victorian and hybrid NSW. DBHs were 139 & 122 mm for *E. nitens* from Victoria & NSW.

The frequency distributions of hybrid DBHs were bimodal, with peaks for the 'runts' and for the vigorous trees, showing clearly the few very vigorous trees and many runts in a family. The possibilities of clonal forestry with the hybrids were exciting, following the Mondi success in South Africa. It was clear, however, early on, that the hybrids were not cold hardy enough for most NI eucalypt sites or for the SI. The problems of cutting propagation in *E. nitens* itself were off-putting as this species was difficult to propagate from coppice, which with *E. grandis* gives rejuvenation of a clone (Shelbourne et al. 1999b).

This attempt at inter-species hybrids was a good introduction to the related problems, and unfortunately the characteristics of *E. nitens* and other ash species did not promise as well as the popular *E. grandis x E. urophylla* hybrids and others in the sub-tropics of South Africa and Brazil. Later assessment data from our hybrids confirmed the early results.

**R. 2709. R.P. Kibblewhite and C.J. McKenzie 1999: Kraft fibre variation among 9 trees of 15-year-old *Eucalyptus fastigata* and comparison with *E. nitens*. Appita J. 52(3): 218–225.**
Variation in wood basic density and chemistry, fibre and handsheet properties of 29 trees of *E. fastigata* age 15 were compared with 29 trees of *E. nitens*. *E. fastigata* had higher lignin content and lower pulp yield with wider ranges of fibre collapse and handsheet bulk than *E. nitens*. The high end of the bulk range (low fibre collapse) was similar, but only 20% of the trees had acceptably high bulk. *E. fastigata* was generally less desirable than *E. nitens* on account of its high lignin and lower pulp yield. The *E. fastigata* study was the same design as the *E. nitens* one. This study links with another on *E. regnans* (even lower density and bulk), and others on bulked samples of *E. globulus*, *E. maidenii* and *E. nitens*. The bottom line was that none of 'our' eucalypts was anything like as good for kraft pulping as *E. globulus*, which doesn't grow well in NZ (Kibblewhite & McKenzie 1999).

***Comment*** Good research (see also the mechanical pulping studies) in which the genetic sampling influence and material from GTI was important.

**R.P. Kibblewhite, M.J.C. Riddell and C.J.A. Shelbourne 2000a: Variation in wood, kraft fibre, and handsheet properties among 29 trees of *Eucalyptus regnans*, and comparison with *E. nitens* and *E. fastigata*. NZJFS 30.**
Variation in and relationships between wood, chemical, kraft fibre, and kraft handsheet properties were studied in 29 individual trees of *Eucalyptus regnans* F. Mueller, aged 20–21 years, grown in Kaingaroa Forest. Means and ranges of these characteristics for *E. regnans* were also compared with those for similar 29-tree samples from *E. nitens* (Deane et Maiden) Maiden and *E. fastigata* Deane et Maiden from the same forest, aged 15–16 years.

The trees sampled of *E. regnans* were of lower wood density (in spite of their greater age) and had kraft fibres that were longer, broader, thinner-walled, and had higher levels of collapse than those of either of the other species. *Eucalyptus regnans* individual-tree pulps showed the widest range of apparent sheet density compared with the other two species but had a similar mean, which was well predicted by wood density or by the level of fibre collapse in handsheets. *Eucalyptus regnans* trees had a good pulp yield, similar to those of *E. nitens*. However, *E. regnans* grown for 15 years could be expected to give a kraft pulp very deficient in handsheet bulk (Kibblewhite et al. 2000a).

***Comment*** This study of 29 trees of *E. regnans* links with the other 29-tree studies of *E. nitens* and *E. fastigata*. These were all of mature age for pulpwood (or somewhat over mature for *E. regnans)* and show up the deficiency in handsheet bulk (inverse of apparent density) of all 3 species, especially *E. regnans,* which is well correlated with wood density. This research on individual tree variation in kraft fibre and handheet properties was enormously expensive and apparently quite unique. My role was mainly in planning the sampling and in assisting in the write up.

**R.P. Kibblewhite, B.L. Johnson and C.J.A. Shelbourne 2000b: Kraft pulp qualities of *Eucalyptus nitens, E. globulus* and *E. maidenii* at ages 8 and 11 years. NZJFS 30 (3): 347–357**

Kraft fibre and pulp properties were assessed for ten-tree bulked chip samples from 8- and 11-year-old species/provenance trials of *Eucalyptus globulus* Labill. *E. nitens* (Deane et Maiden) Maiden and *E. maidenii* Labill, grown on two sites south of Kaikohe in Northland. Mean basic density of bulked chip samples ranged from 447 kg/m$^3$ for both ages of *E. nitens* to 576 kg/m$^3$ for 11-year-old *E. maidenii*. Pulp yields for all wood types were similar, from 54.5% to 55.6%.

The kraft fibres of *E. maidenii* were somewhat longer, with higher wall area (coarseness) than those of *E. globulus*, which were in turn of higher coarseness than those of *E. nitens*. Fibre collapse potential (as indicated by fibre width/thickness ratio) of both *E. maidenii* pulps and the *E. globulus* 11-year-old material was much less than that for *E. nitens*. For these six wood origins, pulps of premium quality were obtained from *E. globulus* aged 11 and from *E. maidenii* aged 8 years. The pulp from 11-year-old *E. maidenii* was too high in bulk, requiring excessive refining, and the pulps from *E. nitens* (aged 8 and 11 years) were deficient in bulk and unsuitable for many eucalypt market kraft end uses (Kibblewhite et al. 2000b).

**R.B. McKinley, C.J.A. Shelbourne, J.M. Harris and G.D. Young 2000: Variation in whole-tree basic wood density for a range of plantation species grown in New Zealand. NZJFS 30 (3): 436–446.**

Whole-tree basic wood density of individual trees of a variety of species has been determined in many different studies in New Zealand since 1977. These data were recently collated, and whole-tree values for each species have been aggregated across sites into 5-year age-classes, from <7 years up to 70 years. Means, ranges and standard deviations of each species/age-class have been tabulated for a total of 968 trees and 13 species or species groups, with numbers of trees/species varying from 15 to 232. Sites sampled were mainly in the central and northern parts of the North Island (latitude 39°30′–35°25’S). Logarithmic regression equations fitted to the age-class mean densities for each species have provided predictions of whole-tree basic density with age and, in conjunction with predicted volume yields, were used in a related study to predict stem dry matter production per hectare for different species.

The species included were *Acacia dealbata*, *A. mearnsii*, cypresses (data from *Chamaecyparis lawsoniana*, *Ch. nootkatensis* x *Cupressus macrocarpa*, *C. lusitanica*, *Pseudotsuga menziesii*, *Eucalyptus fastigata*, *E. globoidea*, *E. globulus*, *E. maidenii*, *E. muelleriana*, *E. nitens*, *E. pilularis*, *E. regnans* and *E. saligna*. Predicted mean whole-tree basic density (from logarithmic regression) for the eucalypts (30-year-old trees) varied from 452 kg/m$^3$ for *E. regnans* to 623 kg/m$^3$ for *E. globoidea*.

No data at age 30 years were available for *Acacia mearnsii* and *E. maidenii*, but their mean density was, respectively, 658 kg/m$^3$ at age 14 and 572 kg/m$^3$ at age 11 years. Whole-tree mean density was about 418 kg/m$^3$ for the cypresses and 406 kg/m$^3$ for *Ps. menziesii*, almost irrespective of tree age. A notable feature of these data was the great variability in density between trees in a stand, but variation in whole-tree density with age showed consistent patterns for different species, in spite of confounded effects of site and stocking.

***Comment*** Useful summary of large amount of data. Important reference for alternative spp. review

**C.J.A. Shelbourne, B.T. Bulloch, R.L. Cameron and C.B. Low 2000: Results of provenance testing *Acacia dealbata, A. mearnsii* and other Acacias at ages 7 and 5 years in New Zealand. NZJFS 30 (3): 401–421.**

Provenance and species trials of *Acacia dealbata* (31 seedlots) and *A. mearnsii* (23 seedlots), together with 1 or a few seedlots each of *A. decurrens*, *A. falciformis*, *A. filicifolia*, *A. melanoxylon* and *A. silvestris*, were planted at five sites in 1989 and at two sites in 1991. All trials were assessed in 1996 at ages 7 years and 5 years (for 1991-planted trials) for diameter at breast height (DBH), bole straightness (score 1–9), and malformation (score 1–5). A small subsample of trees from seven seedlots each of *A. mearnsii* (from two sites) and of *A. dealbata* (from three sites) were assessed for height, DBH and wood density (5 mm. pith-to-bark cores) at age 7 years.

*Acacia dealbata* and *A. mearnsii* greatly outgrew the other species and were of much better form than all except *A. silvestris*. Both *A. dealbata* and *A. mearnsii* showed large and significant differences between seedlots (mostly provenances) in diameter growth and bole straightness, and *A. dealbata* did so in malformation.

For *A. dealbata* trials aged 7 years, overall mean DBH varied from 206 mm at Kinleith to 145 m at Tuki Tuki (Havelock North); for *A. mearnsii* (not assessed at Kinleith), site mean DBH ranged from 151 mm at Tuki Tuki to 135 mm at Pohangina (Palmerston North). Averaged across three sites, DBH of the best three seedlots of *A. dealbata* was 175 mm.compared with 162 mm for the best three seedlots of *A. mearnsii*.

The seedlots of *A. dealbata* with better growth and form were reasonably free of malformation and their bole straightness was acceptable for sawlogs. All seedlots of *A. mearnsii* were much more sinuous in stem form than *A. dealbata*, such that very few trees would be straight enough for sawlogs.

Height of *A. dealbata* averaged 12 m at Tuki Tuki and up to 15 m at Kinleith. Height of *A. mearnsii* averaged 12 m at Tuki Tuki and Emerald Hills. Basic wood density of *A. dealbata* averaged 369 kg/m$^3$ at Kinleith, 371 kg/m$^3$ at Emerald Hills and 412 kg/m$^3$ at Tuki Tuki. In contrast, wood density of *A. mearnsii* averaged 553 kg/m$^3$ at Tuki Tuki and 556 kg/m$^3$ at Emerald Hills. The better provenances of *A. dealbata* showed promise for sawlog and pulpwood production, based on rapid growth rate, acceptable bole straightness and wood density and (from other work) good sawing, seasoning and appearance characteristics. The rapid growth rates and high wood density of *A. mearnsii* in New Zealand, as well as existing market acceptance of South African material for pulp, indicate a real potential for this species for pulpwood.

***Comment*** Unfortunately, the good early performance of A. dealbata and A. mearnsii was not maintained at later ages. The interest in acacias soon waned.

**C.J.A. Shelbourne, C.B. Low and P.J. Smale 2000: Eucalypts for Northland: seven- to 11-year results from trials of nine species at four sites. NZJFS 30 (3): 366–383.**

Species and provenance trials of eucalypts planted in Northland between 1988 and 1993 included several provenances each of *Eucalyptus fastigata. E. regnans. E. saligna, E. botryoides, E. grandis, E. nitens, E. globulus*, and *E. maidenii*. Trials were all located between Kaikohe and Dargaville (latitudes 35°31′ to 35°48′), with two trials at Carnation Road and two at Walker Road (aged 11 and 9 years) and one of *E. nitens* only at Karaka Road. Trial designs were mainly 64-tree square plots with 2–5 replicates. Trees were assessed for breast height diameter over-bark, bole straightness, malformation and crown health. Results were expressed as provenance and species means at each site, as basal area per hectare over-bark, volume per hectare under-bark and frequencies of crop trees for mortality and runts (suppressed sub-dominants).

*Eucalyptus fastigata* at its single test site showed best growth and health of all species (mean annual increment (MAI) at Carnation Road averaged 52 m$^3$/ha) but suffered some basal and upper stem forking. *Eucalyptus regnans* averaged 50 m$^3$/ha/year, with good crown health. *E. nitens* of central Victorian provenances showed poor crown health and high mortality despite good earlier growth. *Eucalyptus saligna* (and *E. grandis*) grew more slowly than other species and showed a high proportion of 'runts' (small DBH and frequent malformation). *Eucalyptus maidenii*, planted in only one subsidiary trial at Carnation Road (and at Knudsen Road), had better crown health, higher survival and better growth than *E. saligna, E. grandis* and *E. globulus*, though its volume growth appeared to be less than *E. fastigata* in the main trial.

*Eucalyptus nitens* of central Victorian provenances was evidently poorly adapted and unlikely to continue its earlier good growth, and even the healthier NSW provenances appeared insecure. *E. fastigata* was a winner for growth and health, closely followed by *E. regnans. E. globulus* had generally poor health and slower growth than *E. nitens*. *Eucalyptus maidenii*, although a slower starter, had good crown health and good survival and showed higher wood density (from other studies) than *E. nitens* and *E. globulus*, and by inference, *E. fastigata* and *E. regnans*. *Eucalyptus saligna*, originally the preferred species in Northland, produced much less volume than the other species. *Eucalyptus botryoides* failed completely due to early possum damage.

***Comment*** This work is from trials almost old enough to form a valid basis for evaluating these species in Northland sites and an example of the necessity of obtaining data from 10- to 15-year-old trials in eucalypt improvement work.

**R. 2813. R.B. McKinley, C.J.A. Shelbourne, C.B. Low, B. Penellum and M.O. Kimberley 2002: Wood properties of young *Eucalyptus nitens, E. globulus* and *E. maidenii* in Northland, New Zealand. NZJFS 32(3): 334–356.**
This was an in-depth study of young (eight years and 11 years) *E. nitens, E. globulus* and *E. maidenii*, ancillary to a kraft pulping study. Ten trees of each species and of each age at, respectively, Knudsen and Carnation roads, near Kaikohe, latitude 35° 30′, altitude ca. 200 m, 1800–1900 mm rainfall, were selected to compare wood, pulping and some lumber properties. The sample trees were selected across the species' range (determined first by core sampling for density) to approximate the species mean. Wood properties were measured on discs cut from the base and at 5 m intervals up the tree. A 1 metre billet was cut between 5 and 6 m in height and used for measuring MOE on test sticks.

Mean whole tree densities were 574, 540 and 451 kg/m$^3$ for *E. maidenii, E. globulus* and *E. nitens*. Density increased with height in *E. nitens* and *E. globulus* and decreased in *E. maidenii*. Internal checking at height 6 m was severe for *E. nitens*, occasional in *E. globulus* and completely absent in *E. maidenii*. The continued good growth and health of *E. maidenii* compared with very poor health of *E. globulus* and *E. nitens*, and its lack of checking, low spiral grain, low shrinkage and collapse and excellent strength and stiffness promised well for solid wood products (McKinley et al. 2002).

***Comment*** *E. nitens* was unsuited to Northland low altitude, warm sites, based on its native 600 m altitude habitat in Victoria. *E. globulus* had disappointed because of disease and possibly general maladaptation to the higher rainfall climates of NZ.

**C.J.A. Shelbourne, I.D. Nicholas, R.B. McKinley, C.B. Low, R.M. McConnochie and M.J.F. Lausberg 2002a, b: Wood density and internal checking of young *Eucalyptus nitens* in New Zealand as affected by site and height up the tree. NZJFS 32.**
Whole-tree basic density and internal checking were assessed in *Eucalyptus nitens* (Deane et Maiden) Maiden by sampling 15 trees/site, each involving the same seedlot of Victorian provenance grown at a stocking of 1111 stems/ha. Six New Zealand sites, planted the same year, four in the North Island and two in the South Island were sampled. Sites ranged in altitude from 40 to 540 m and in latitude from 35°52'S (Dargaville) to 45°55'S (Drumfern). Site mean whole-tree density ranged from

428 kg/m$^3$at Raweka (Whakatane) to 476 kg/m$^3$ at Mangakahia (Dargaville). Density at Kinleith, Wainui (both central North Island), and Millers Flat and Drumfern (southern South Island) varied little, from 445 kg/m$^3$ to 459 kg/m$^3$. From these and previous results, there was some indication that very high rainfall and high levels of foliar nitrogen, phosphorus and magnesium led to extremely low wood density. Whole-tree density increases with age and increases with height up the stem (Shelbourne et al. 2000a).

Internal checking, assessed in one breast-height disc per tree, was prevalent at all sites, especially in outer heartwood in both slowly kiln-dried and air-dried samples. More checks were found in air-dried discs than in kiln-dried. Many more checks were found at the North Island sites, Mangakahia, Raweka and Kinleith, than at the high-altitude central plateau site, Wainui, or the South Island sites. High numbers of checked rings and total checks were associated with higher mean annual temperatures, short green crowns and poor crown health. Far fewer checks were found at cooler sites where trees had much better crown health and longer green crowns (Shelbourne et al. 2000b).

Checking data from an earlier study numbers of checks varied enormously among trees. Importantly, checking fell to zero above height 11.4 m. Numbers of checks in the board cross-section correlated moderately with those in the breast-height disc. Excessive checking in *E. nitens* will reduce its potential for utilisation for appearance-grade lumber, particularly where crown health is poor.

**C.J.A. Shelbourne, B.T. Bulloch, C.B. Low and R.M. McConnochie 2002c: Performance to age 22 years of 49 eucalypt species in the Wairarapa district, New Zealand, and results from other trials. NZJFS 32(2): 256–278.**

Trials of 49 eucalypt species were established in 1979 in the Wairarapa district at Kahuiti and Pakaraka, New Zealand, originally to test species for their potential to stabilize erodable land for pastoral use. Trials were planted in a randomized complete block design with five replications of four-tree row plots of each seedlot (paired rows of four trees of species with only a single seedlot).

The species included *Corymbia maculata [E. maculata]*, *E. cladocalyx*, 4 stringybarks (including *E. muelleriana* and *E. globoidea*), 9 ashes (including *E. fastigata, E. regnans* and *E. obliqua*), 7 peppermints, and 18 gums (including *E. nitens*). Because of heavy thinning at Pakaraka, the Kahuiti trial only was assessed at age 22 years for diameter at breast height (DBH), stem straightness, malformation, crown health and number of potential 5 m sawlogs per tree.

The 12 best-grown species at Kahuiti were ranked as follows: *E. globoidea, E. muelleriana* (stringybarks), *E. obliqua, E. fraxinoides, E. regnans* (ashes), *E. cordata* (gum), *E. delegatensis, E. fastigata, E. sieberi* (ashes), *E. cinerea, E. kartzoffiana* and *E. nitens* (gums). The ashes, with addition of *E. nitens*, showed a combination of best diameter growth (apart from the two stringybarks), straightest stems, least malformation, good crown health and largest number of sawlogs per tree of all groups.

The peppermints were generally slower growing and more sinuous than the ashes. Some of the gums grew well and most survived better on this adverse, eroded site than the other groups. Superiority of *E. fastigata* and *E. obliqua* was confirmed by other trials in Hawke's Bay and the Wairarapa region. The apparent good growth and health of *E. globoidea* and *E. muelleriana* in the Wairarapa district, also reported in trials in Northland, hint at the potential of these known good saw-timber species (Shelbourne et al. 2002c).

***Comment*** This is our biggest species trial and could well have been planted on other sites where it would have given us a 'global' look at species choice.

**H.M. McKenzie, J.C.P. Turner and C.J.A. Shelbourne 2003a, b: Processing young plantation-grown *Eucalyptus nitens* for solid-wood products. 1. Individual-tree variation in quality and recovery of appearance-grade lumber and veneer. NZJFS 33(1): 62–78.**

A stand at Golden Downs (New Zealand northern SI) of *Eucalyptus nitens* (Deane and Maiden) Maiden was pruned up to height 8 m and grown for 15 years at low stocking to 57 cm diameter at breast height. This stand provided 15 trees, preselected for a range of wood density. Lumber and sliced veneer were cut from the 5 m butt logs, veneer was rotary-peeled from the second logs from height 7 to 13 m, and each tree was evaluated for production of appearance-grade lumber and veneer (Mckenzie et al. 2003a).

Butt-log quality was good, as pruning had effectively restricted the knotty core, and there was little decay from branches in either butt logs or veneer billets. Longitudinal growth stresses varied widely among trees, resulting in log end-splitting and sawlog flitch movement during sawing (spring), which led to crook in sawn timber, substantially reducing timber conversion in some trees. Collapse and internal checking were prevalent in air-dried lumber, and numbers of checks varied widely among trees. Face-checking was found in boards from all trees after kiln-drying and reconditioning, and even those with very few face checks.

Veneer thickness varied unacceptably, caused probably by incorrect knife-and pressure-bar settings. Veneer splitting also varied among trees, and was worse in butt-log than in second-log veneers. Knots severely downgraded structural plywood veneer grades, <8% of sheets from the second logs being acceptable compared with 87% of sheets from the pruned butt logs. Stiffness of veneer sheets was successfully measured using a sonic device (Pundit) to sort veneers for manufacture of laminated veneer lumber (McKenzie et al. 2003b).

**H.M. McKenzie, C.J.A. Shelbourne, M.O. Kimberley, R.B. McKinley and R.A.J. Britton 2003c: Processing young plantation-grown *Eucalyptus nitens* for solid-wood products. 2. Predicting product quality from tree, increment core, disc and 1 m billet properties. NZJFS 33(1): 79–113.**

Butt logs of 15 trees of *Eucalyptus nitens* (Dean et Maiden) Maiden, aged 15 years, were cut into appearance-grade lumber and sliced veneer, and the second logs into rotary-peeled veneer. A 1-m. billet was removed from between butt and second logs of each tree, as well as discs at successive heights. In addition, breast-height increment cores and breast-height measurements of longitudinal growth strain served to characterise the wood properties and processing, product and clearwood mechanical properties of each tree.

Fibre dimensions, density and microfibril angle were measured by SilviScan on a sample from height 6 m. Boards were quarter-sawn from the 1 m billet and air- and dehumidifier-dried, and internal checking and shrinkage were measured on these boards and on discs from height 6 m.

"Sawability" variables of the butt log (log-end splits, flitch movement off the saw, timber crook and timber conversion percentage) showed strong intercorrelations with one another and with longitudinal growth strain, measured at breast height on the standing tree. Amounts of internal checking and collapse in the air-and then kiln-dried butt-log boards were strongly correlated with checking, measured on discs and on the billet boards. Shrinkage of the 1 m boards and of blocks from the 6 m height disc was correlated moderately with collapse and checking in the butt-log boards.

Clearwood modulus of elasticity measured on eight test sticks were cut from the billet from height 6–7 m. There was a strong increasing gradient in MOE from pith to bark as well as wide variation among trees. Density showed only a small pith to bark increase, while microfibril angle rapidly decreased from the pith over the first seven rings.

Clearwood modulus of elasticity was strongly correlated with the SilviScan density/microfibril angle ratio, as was modulus of elasticity of individual test sticks.

Trees varied strongly in product characteristics and wood properties, and there were strong correlations (a) between breast-height growth strain and sawability characteristics, and (b) between checking and collapse in butt-log boards, and tangential shrinkage and checking measured on discs. Good correlations of appearance-lumber properties with similar traits measured on the standing tree or from cores, discs and a 1 m billet indicated that effective evaluation of trees is possible without recourse to full-scale sawing studies. (McKenzie et al. 2003c).

**L.D. Gea, R.M. McConnochie and S. Wynyard 2007: Parental reconstruction for breeding, deployment and seed orchard management of *Eucalyptus nitens*. NZJFS 37 (1): 23–36.**

A new open-pollinated breeding strategy for *Eucalyptus nitens* (Deane & Maiden) Maiden in New Zealand was explored using microsatellite markers to reveal the male parental identity of forwards selections. Microsatellites are the preferred markers to reveal genetic relationships between individuals, largely owing to their co-dominant inheritance. Forwards selection of individuals for the breeding population and for future deployment was simulated using ten open-pollinated seedling offspring from each of ten clones in a clonal seed orchard. A set of 15 microsatellite markers was chosen from the 41 initially tested. Ninety of the 100 progeny sampled matched consistently to a single mother and father and 13 of these were evidently selfs. Eight had a maternal match only; this would indicate that either there was contamination by pollen from outside the orchard or there was an occasional failure by the marker set to identify the orchard pollen. One seedling had no maternal match and it was not possible to discriminate between two fathers in in another (Gea et al. 2007).

There was a broad level of outcrossing at the individual and provenance levels, and there appears to be little indication that individual clones favour specific pollens.

Estimates of the coefficient of inbreeding and coefficient of co-ancestry were derived for the seed orchard and forwards selections.

**T.G. Jones, R.M. McConochie, C.J.A. Shelbourne and C.B. Low 2010: Sawing and grade recovery of 25-year-old *Eucalyptus fastigata, E. globoidea, E. muelleriana* and *E. pilularis*. NZJFS 40: 19–31.**
The processing characteristics of 25-year-old *Eucalyptus fastigata, E. globoidea, E. muelleriana* and *E. pilularis* from Rotoehu Forest, New Zealand, were evaluated to determine if these species could be used to produce high-quality timber on shorter rotations.

The butt- and second-logs of 15 trees of each species were quarter-sawn and flat-sawn, respectively, and the boards assessed for shrinkage and distortion, visual and mechanical properties and surface hardness.

Growth-stress release during sawing, combined with end-checking during drying, resulted in board end-splitting that reduced the sawn recovery in *E. fastigata* compared with the other species, and produced high levels of crook in the quarter-sawn boards of all species. There was no surface checking and little collapse and internal checking on drying.

The proportions of visual clears and No.1 cutting grades were low, particularly for *E. muelleriana* and *E. pilularis*, due to the presence of knots. The boards of all species had high values of density, modulus of elasticity and surface hardness, and machine stress grades of MSG10 to MSG15. These species have the potential to produce high-quality timber on 25-year rotations, but pruning will be required to improve visual grades so that a higher proportion of boards can be used in appearance applications (Jones et al. 2010).

## References


10. G.B. Sweet 1962: The future supply of high-quality eucalypt seed for New Zealand. An analysis of the problems involved.

R. 40. I.J. Thulin and T. Faulds 1966?: Grafting of eucalypts. NZ J. For. 8(4):664–667.

R. 1379. M.D. Wilcox 1980: Genetic improvement of eucalypts in New Zealand. NZJFS 10(2): 3–359.

R. 1295. M.D. Wilcox 1979: The ash group of eucalypts. NZJFS 9(2): 133–144.

R. 1298. M.D. Wilcox and I.J. Thulin 1979: Growth of *Eucalyptus regnans* in a plot at Rotorua. NZJF 9 (2): 166–169.

R. 1307. M.D. Wilcox, T. Faulds, T.G. Vincent and B.R. Poole *Eucalyptus regnans* F.Muell. Aust. For. Res. 10:169–84.

R. 1308. D.A. Rook, M.D. Wilcox, D.G. Holden and I.J. Warrington 1980: Provenance variation in frost tolerance of *Eucalyptus regnans* F. Muell. Aust. For. Res. 10:213–238.

M.I. Menzies, D.G. Holden, D.A. Rook and A.K. Hardacre 1981: Seasonal frost-tolerance of *Eucalyptus saligna, E. regnans* and *E. fastigata.* NZJFS (11).

M. Dick:1982 Leaf-inhabiting fungi of eucalypts in New Zealand. NZJFS (12).

R. 1503. M.D. Wilcox 1982a: Anthocyanin polymorphism in seedlings of *Eucalyptus fastigata* Deane et Maid. Aust. For. Res. 30: 501–9.

R. 1604. M.D. Wilcox, T. Faulds and T.G. Vincent 1980a: Genetic improvement of *Eucalyptus saligna* Sm in New Zealand. IUFRO Symposium and Workshop, Aguas de Sao Pedro, Sao Paulo, Brazil Aug. 1980 (also published 1982 by FAO Rome in Forest Genetic Resources Information No 11).

M.D. Wilcox 1982b: Genetic variation in frost tolerance, early height growth rate and incidence of forking among and within provenances of *E. fastigata*. NZJFS (12).

R. 1605. M.D. Wilcox, D.A. Rook and D.G. Holden 1980b: Provenance variation in frost resistance of *Eucalyptus fastigata* Deane & Maid. IUFRO Symposium and Workshop, Aguas de Sao Pedro, Sao Paulo, Brazil. Aug. 1980. (also published 1982 by FAO Rome in Forest Genetic Resources Information No 11).

R. 1590. M.D. Wilcox 1982c: Preliminary selection of suitable provenances of *Eucalyptus regnans* for New Zealand. NZJFS 12(3): 468–479.

R. 1591. M.D. Wilcox 1982d: Selection of genetically-superior *Eucalyptus regnans* using family tests. NZJFS 12(3): 480–493.

FRI Bulletin No. 95. M.D. Wilcox, J.T. Miller, I.M. Williams and D.W. Guild 1985: *Eucalyptus* species trials in Longwood forest, Southland. Bulletin No. 95, FRI, NZFS, PB. Rotorua, NZ.

J.N. King and M.D. Wilcox 1988: Family tests as a basis for the genetic improvement of *Eucalyptus nitens* in New Zealand. NZJFS 18.

R.L. Hathaway and M. King 1986: Selection of *Eucalyptus* species for soil conservation planting in seasonally dry hill country. NZJFS 16.

R. 2053. M.D. Wilcox and R.L. Hathaway 1988: Use of Australian trees in New Zealand. Proc. AFDI International Forestry Conference for the Australian Bicentenary 1988, Albury, New South Wales, Australia Volume III.

R. 2467. P.G. Cannon and C.J.A. Shelbourne 1991: The New Zealand eucalypt breeding programme. IUFRO Symposium on Intensive Forestry: The Role of Eucalypts P2.0201 Productivity of eucalypts. Durban, South Africa 2–6 September, 1991.

R. 2483. P.G. Cannon and C.J.A. Shelbourne 1993: Forward selection plots in breeding programmes with insect-pollinated tree species. NZJFS 23(1): 3–9.

M.J.F. Lausberg, K.F. Gilchrist and J.H. Skipwith 1995: Wood properties of *Eucalyptus nitens* grown in New Zealand. NZJFS (25).

L.D. Gea, R.M. McConnochie and N.M.G. Borralho 1997: Genetic parameters for growth and wood density traits in *Eucalyptus nitens* in New Zealand. NZJFS (27).

R. 2659. R.P. Kibblewhite, M.J.C. Riddell and C.J.A. Shelbourne 1998: Kraft fibre and pulp qualities of 29 trees of New Zealand grown *Eucalyptus nitens*. Appita 51(2): 114–121.

C.B. Low and C.J.A. Shelbourne 1999: Performance of *Eucalyptus globulus, E. maidenii, E. nitens*, and other eucalypts in Northland and Hawke's Bay at ages 7 and 11 years. NZJFS 29 (2): 274–288.

C.J.A. Shelbourne, S.O. Hong, R.M. McConnochie and B. Pierce 1999a: Early results from trials of interspecific hybrids of *Eucalyptus grandis* with *E. nitens* in New Zealand. NZJFS 29(2): 251–262.

R. 2709. R.P. Kibblewhite and C.J. McKenzie 1999: Kraft fibre variation among 29 trees of 15-year-old *Eucalyptus fastigata* and comparison with *E. nitens*. Appita J. 52(3): 218–225.

R.P. Kibblewhite, M.J.C. Riddell and C.J.A. Shelbourne 2000a: Variation in wood, kraft fibre, and handsheet properties among 29 trees of *Eucalyptus regnans*, and comparison with *E. nitens* and *E. fastigata*. NZJFS 30.

R.P. Kibblewhite, B.L. Johnson and C.J.A. Shelbourne 2000b: Kraft pulp qualities of *Eucalyptus nitens, E. globulus* and *E. maidenii* at ages 8 and 11 years. NZJFS 30 (3): 347–357

C.J.A. Shelbourne, S.O. Hong, R.M. McConnochie and B. Pierce 1999b: Early results from trials of interspecific hybrids of *Eucalyptus grandis* with *E. nitens* in New Zealand. NZJFS 29(2): 251–262.

R.B. McKinley, C.J.A. Shelbourne, J.M. Harris and G.D. Young 2000: Variation in whole-tree basic wood density for a range of plantation species grown in New Zealand. NZJFS 30 (3): 436–446.

C.J.A. Shelbourne, B.T. Bulloch, R.L. Cameron and C.B. Low 2000a: Results of provenance testing *Acacia dealbata, A. mearnsii* and other Acacias at ages 7 and 5 years in New Zealand. NZJFS 30 (3): 401–421.

C.J.A. Shelbourne, C.B. Low and P.J. Smale 2000b: Eucalypts for Northland: 7- to 11-year results from trials of nine species at four sites. NZJFS 30 (3): 366–383.

R. 2813. R.B. McKinley, C.J.A. Shelbourne, C.B. Low, B. Penellum and M.O. Kimberley 2002: Wood properties of young *Eucalyptus nitens, E. globulus* and *E. maidenii* in Northland, New Zealand. NZJFS 32(3):334–356.

C.J.A. Shelbourne, I.D. Nicholas, R.B. McKinley, C.B. Low, R.M. McConnochie and M.J.F. Lausberg 2002a: Wood density and internal checking of young *Eucalyptus nitens* in New Zealand as affected by site and height up the tree. NZJFS 32.

C.J.A. Shelbourne, I.D. Nicholas, R.B. McKinley, C.B. Low, R.M. McConnochie and M.J.F. Lausberg 2002b: Wood density and internal checking of young *Eucalyptus nitens* in New Zealand as affected by site and height up the tree. NZJFS 32.

C.J.A. Shelbourne, B.T. Bulloch, C.B. Low and R.M. McConnochie 2002c: Performance to age 22 years of 49 eucalypt species in the Wairarapa district, New Zealand, and results from other trials. NZJFS 32(2): 256–278.

H.M. McKenzie, J.C.P. Turner and C.J.A. Shelbourne 2003a: Processing young plantation-grown *Eucalyptus nitens* for solid-wood products. 1. Individual-tree variation in quality and recovery of appearance-grade lumber and veneer. NZJFS 33(1): 62–78.

H.M. McKenzie, C.J.A. Shelbourne, M.O. Kimberley, R.B. McKinley and R.A.J. Britton 2003b: Processing young plantation-grown *Eucalyptus nitens* for solid-wood products. 2. Predicting product quality from tree, increment core, disc and 1 m. billet properties. NZJFS 33(1): 79–113.

H.M. McKenzie, C.J.A. Shelbourne, M.O. Kimberley, R.B. McKinley and R.A.J. Britton 2003c: Processing young plantation-grown *Eucalyptus nitens* for solid-wood products. 2. Predicting product quality from tree, increment core, disc and 1 m. billet properties. NZJFS 33(1): 79–113.

L.D. Gea, R.M. McConnochie and S. Wynyard 2007: Parental reconstruction for breeding, deployment and seed orchard management of *Eucalyptus nitens*. NZJFS 37 ( 1): 23–36

T.G. Jones, R.M. McConochie, C.J.A. Shelbourne and C.B. Low 2010: Sawing and grade recovery of 25-year-old *Eucalyptus fastigata, E. globoidea, E. muelleriana* and *E. pilularis*. NZJFS 40: 19–31.

# Chapter 9
# Finale Eucalypts

Eucalypt improvement in NZ was tackled across a whole genus, comprising hundreds of species, from 1975 to 1980. Based on common knowledge, on expert experience and on the inspiration of Mike Wilcox, seven species were chosen in which to start breeding programmes: *Eucalyptus botryoides*, *E. saligna*, *E. regnans*, *E. delegatensis*, *E. fastigata*, *E. obliqua* and *E. nitens*.

Other experts had nominated other species, mainly for their sawn timber properties. Harry Bunn chose *E. muelleriana*, *E. pilularis*, *E. botryoides*, *E. saligna*, *E. scabra* (*globoidea*), *E. obliqua*, *E. delegatensis*, *E. regnans* and *E. fastigata*, which included several known good sawn timber species. Mike Wilcox's list included *E. regnans* and *E. fastigata* which were being planted for pulp by New Zealand Forest Products at that time. There was little information about variation in pulping properties between and within species then.

Mike Wilcox was the main initial source of planning and field management of the various breeding programmes. He outlined the basic breeding strategy for eucalypts, which was to introduce seed of individual trees of several native provenances from Australia and plant these open-pollinated families as the basis of a breeding population, with some selection and open-pollinated (OP) seed collection in NZ stands. Little was known in NZ about the different species apart from their appearance, growth rate and health. The best trees from the most promising provenances in the BP (breeding population) were then selected, grafted and planted in clonal seed orchards.

Seed collections from these selections could then be planted in seedling seed orchards or seed stands. Of the seven species listed above, only *E. fastigata*, *E. nitens* (in the SI only) and just possibly *E. regnans* are still in breeding programmes in 2016. Results of pulping research later indicated *E. regnans*, *E. fastigata* and *E. nitens* are inferior for kraft pulping (too low in 'bulk').

A large group of 24 species were also listed as 'under test' in species trials, including 7 ashes, karri from WA, 2 stringybarks, blackbutt and *E. pyrocarpa*, and the rest gums. Few of these species were properly tested in species trials. Individual breeding programmes in 1980 included:



C. J. A. Shelbourne, M. Carson, *Tree Breeding and Genetics in New Zealand*,
https://doi.org/10.1007/978-3-030-18460-5_9

*E. saligna* (43 native, 55 NZ families) planted in 1976 and clonal orchards planted in 1976
*E. regnans* (144 families, native and NZ)
*E. delegatensis* (38 native, 52 NZ families, planted in 1977 but later rejected, too slow)
*E. fastigata* (48 NZ, 73 native families and 6 South Africa families planted 1979
*E. nitens* (80 native families of only Victorian provenances) planted in 1979.

This was an incredible effort over a short period from 1974 to 1980. Five provenance-progeny trials with 100–250 families each of five species were planted at several sites. The genetic improvement program for *E. regnans* was started in 1977 with 144 families. Thirty-six bulked provenance lots were also tested at Wiltsdown (Kinleith) and Kaingaroa. Frost damage was evident 1 year from planting and varied widely among families.

It was believed OP families were not true half-sibs so that provenance and family differences were often confounded. Family means are unlikely to provide reliable measures of GCA. However, seed was collected from the best NZ mothers for commercial planting. Family test results were eventually used to thin tests and convert them to SSOs. There seemed little evidence to confine the breeding population to any particular provenance at this stage, and the NZ seedlots were inferior. There was no decision about the provenance composition of the breeding population for *E. regnans* for the CNI.

This result was published early and opened up a lot of the issues in the genetic improvement of eucalypts, such as provenance and family structure, native vs NZ populations and traits for selection.

The *Eucalyptus saligna* programme was started with 78 NZ OP seedlots and 18 Australian provenance seedlots, planted on four sites in 1976. Forty of the NZ seed trees were grafted and small clonal seed orchards planted at Tairua and Kauaeranga. Progeny testing and clonal orchards were probably 'too much, too soon', and this species was insecure in relation to other species. The native provenances generally performed better than NZ progenies. *E. saligna* later performed very poorly relative to *E. fastigata* and *E. regnans* in Northland at age 10 years or more.

A large and comprehensive provenance/progeny trial/breeding population of *E. fastigata* with 126 seedlots was planted in 1979 in a sets-in-replications design at Kinleith and Kaingaroa. The improvement programme for *E. fastigata* was based on 8 Australian native provenances which included 69 OP families and 51 NZ-selected OP families. There were also six seedlots from South Africa. The seedlots varied greatly in frost tolerance, height growth and incidence of forking 1 year after planting. This should prove a sound BP for this promising species.

The Victorian provenance *E. nitens* OP trials at Mt. Tongariro and elsewhere were assessed in 1987 by John King. OP seed collected from resulting selections formed the basis of the second generation of OP families. At the age of 8 years, central Victorian families were 10–15% greater in diameter growth and 35–50% better in tree form scores than seedlots from Eastern Victoria (Errinundra) and southern New South Wales. There were no large or consistent differences among the three provenances of central Victoria – Macalister, Toorongo or Rubicon.

Phil Cannon provided a good round up of the Eucalypt programmes in 1991 for the Durban IUFRO meeting. This was a status report up to 1990 of the various eucalypt species tests, provenance trials and breeding programmes of the first-generation selections from the mid-1970s. Field trials included 7 species trials, 11 provenance trials and 19 progeny tests. *E. regnans* and *E. fastigata* were the best species for altitudes of 300–500 m in the CNI.

*E. nitens* is the best for higher elevations in the NI and in the SI. There were several problems with establishment and health, including weeds, frost, insects, fungi, rabbits and possums. Research on eucalypts decreased heavily during the 1980s, though New Zealand Forest Products continued to plant *E. regnans* and *E. fastigata*. There were 15,000 ha. planted in total.

*E. regnans* breeding included a 300 family second-generation open-pollinated breeding population (OPBP), using single-tree plots on two sites for planting late 1991. *E. nitens* breeding involved 300 OP families of mainly Victorian provenances. Second-generation OP progenies were planted in 1990 in two single-tree-plot trials and one sublined forward selection planting. Heavy *Paropsis* attack occurred on Victorian provenances.

***Comment*** This concentration on Victorian provenances was a mistake, in my view, and so was concentrating test planting on lower altitude NI sites.

*Eucalyptus fastigata* proved to be the best species for resistance to insects and fungal die back but form was not good especially forking. There is plenty of provenance variation as well as family variation in most traits.

*Eucalyptus delegatensis* provenance trials, at two sites at age 8 years, showed Tasmanian provenances with slightly larger diameters than Victorian provenances but poorer in form. Tasmanian provenances had less than half the checking of Victorian provenances. *E. delegatensis*, though an important cold-tolerant species in Southland, grew more slowly and was replaced by *E. nitens*.

Luis Gea analysed data from two open-pollinated second-generation progeny tests of *E. nitens* at age 5 years, including first- and second-generation material of different provenances. Genetic and phenotypic parameters were estimated for diameter at breast height, bole straightness, branching, pilodyn penetration and basic wood density. There were different values for coefficients of relationship (0.25 and 0.5) in native versus plantation progenies, and these needed doubling for native population lots. This indicated full-sib relationship and/or inbreeding in the native progenies.

Paul Kibblewhite in 1998 originated the first individual-tree kraft pulping studies on 29 trees each of *E. nitens*, *E. regnans* and *E. fastigata*, all 15–21 years old, all derived from GTI OP progeny trials. The *E. nitens* study was the first of the pulping studies, with 29 trees of which 9 were also individually mechanically pulped in the pilot plant, an industrial-scale refiner.

There was wide variation in density (390–556 kg/m$^3$) in the 15-year-old trees of Victorian provenances of *E. nitens*. Kraft pulp yield 54–59% varied widely and also kraft pulp properties of fibre dimensions and handsheet properties.

These classic studies explored the whole field of relationships and prediction of pulping and handsheet properties using multiple regression and depended on the variability among individual trees that allowed models to be built.

Variation in wood basic density and chemistry, fibre and handsheet properties of 29 trees of *E. fastigata* at age 15 was compared with *E. nitens*. *E. fastigata* had higher lignin content and lower pulp yield with wider ranges of fibre collapse and handsheet bulk than *E. nitens*. *E. fastigata* was generally less desirable than *E. nitens* on account of its slightly higher lignin and lower pulp yield.

Twenty-nine individual trees of *E. regnans* F. Mueller, aged 20–21 years, grown in Kaingaroa Forest, were evaluated similarly. These were of lower wood density (in spite of their greater age) and had kraft fibres that were longer, broader and thinner-walled and had higher levels of collapse and the widest range of bulk compared with the other two species but had a similar mean. *E. regnans* grown for 15 years was very deficient in handsheet bulk. The bottom line was that none of 'our' eucalypts was anything like as good for kraft pulping as *E. globulus*, which doesn't grow well in NZ.

Another study was originated with 8- and 11-year-old species/provenance trials of *Eucalyptus globulus*, *E. nitens* and *E. maidenii*, grown on two sites south of Kaikohe in Northland. Mean basic density of bulked chip samples ranged from 447 kg/m$^3$ for both ages of *E. nitens* to 576 kg/m$^3$ for 11-year-old *E. maidenii*. Pulp yields for all wood types were similar, from 54.5% to 55.6%. The kraft fibres of *E. maidenii* were somewhat longer, with higher wall area (coarseness) than those of *E. globulus*, which were in turn of higher coarseness than those of *E. nitens*. For these six wood origins, pulps of premium quality were obtained from *E. globulus* aged 11 and from *E. maidenii* aged 8 years.

Unfortunately, the pulp from 11-year-old *E. maidenii,* the only species well adapted to these sites, had bulk which was too high, requiring excessive refining. The pulps from *E. nitens* (aged 8 and 11 years) were deficient in bulk.

Mean whole-tree densities were 574, 540 and 451 kg/m$^3$ for *E. maidenii*, *E. globulus* and *E. nitens*. Density increased with height in *E. nitens* and *E. globulus* and decreased in *E. maidenii*. Internal checking at height 6 m was general and severe for *E. nitens*, occasional in *E. globulus* and completely absent in *E. maidenii*. The continued good growth and health of *E. maidenii* and its lack of checking, low spiral grain, low shrinkage and collapse and excellent strength and stiffness promised well for solid wood products.

Russell McKinley put together whole-tree basic wood density of individual trees of a variety of species, determined in many different studies in New Zealand since 1977. He collated all the data, and whole-tree values for each species have been aggregated across sites into 5-year age classes, from <7 years up to 70 years. Sites sampled were mainly in the central and northern parts of the North Island (latitude 39°30′–35°25′S): useful data for tree breeders.

In 1999 Charlie Low and I assessed species and provenance trials of eucalypts near Kaikohe and at Clive in coastal Hawke's Bay at ages 7 and 11 years, respectively. At Kaikohe, 11 provenances of *E. nitens*, 6 of *E. saligna*, 3 of *E. grandis*, 2 each of *E. globulus* and *E. maidenii* and 1 of *E. robusta* were included. At Kaikohe (age 7 years), best growth was achieved by *E. nitens* (Vic.) and *E. nitens* (NSW) with *E. globulus* and *E. maidenii* equal, but the *E. nitens* (Vic.) provenances showed a widespread progressive loss of the lower crown and eventual death of the tree. The NSW provenances still had healthy crowns. *Eucalyptus globulus* showed a similar problem to *E. nitens* (Vic.), but *E. maidenii* retained good crowns and high survival.

At Clive (age 11 years), a periodically high-water table affected survival and health of some species more than others. *Eucalyptus nitens* of Victorian provenances grew well though survival was poorer than that of the NSW provenance due to crown death. *E. globulus* had grown a little better than both *E. nitens* provenances but showed similar crown death to *E. nitens* (Vic.). *E. maidenii* had a higher mean DBH (219 mm) than all other species and had maintained a high survival. At Patoka, growth, form and health of *E. nitens* (Vic.) and *E. maidenii* were excellent, with *E. maidenii* a little slower-growing.

In 2002 Shelbourne et al. studied whole-tree basic density and internal checking in *Eucalyptus nitens*. We sampled 15 trees/site on 6 sites, each involving the same seedlot of a Victorian provenance planted at 1111 stems/ha., 4 sites in the North Island and 2 in the South Island. Sites ranged in altitude from 40 to 540 m and in latitude from 35°52′S (Dargaville) to 45°55′S (Drumfern). Site mean whole-tree density only ranged from 428 kg/m$^3$ at Raweka (Whakatane) to 476 kg/m$^3$ at Mangakahia (Dargaville). Density at Kinleith, Wainui (both central North Island), and Millers Flat and Drumfern (southern South Island) varied little, from 445 to 459 kg/m$^3$.

Very high rainfall and high levels of foliar nitrogen, phosphorus and magnesium seem to lead to low wood density. Internal checking, assessed in one breast-height disc per tree, was prevalent at all sites, especially in the outer heartwood. Many more checks were found at the North Island sites, Mangakahia, Raweka and Kinleith, than at the high-altitude central plateau site, Wainui, and at the South Island sites. High checking was associated with higher mean annual temperatures, short green crowns and poor crown health. Far fewer checks were found at cooler sites, where trees had much better crown health and longer green crowns.

Bruce Bulloch of the Plant Materials Centre at Palmerston North initiated trials of 49 eucalypt species in 1979 in the Wairarapa District at Kahuiti and Pakaraka, originally to test species for their potential to stabilize erodable land for pastoral use. Trials were planted in a randomized complete block design with five replications of four-tree row plots of each seedlot.

Because of heavy thinning at Pakaraka, the Kahuiti trial only was assessed at age 22 years by Ruth McConnochie on production forestry criteria: diameter at breast height (DBH), stem straightness, malformation, crown health and number of potential 5 m sawlogs per tree. This was 'our' biggest species trial in terms of species coverage, yet it was established by the Plant Materials Centre in Palmerston North!

The 12 best-grown species for DBH ranked *E. globoidea*, *E. muelleriana* (stringybarks), *E. obliqua*, *E. fraxinoides*, *E. regnans* (ashes), *E. cordata* (gum), *E. delegatensis*, *E. fastigata*, *E. sieberi* (ashes), *E. cinerea*, *E. kartzoffiana* and *E. nitens* (gums). The ashes, with addition of *E. nitens*, showed best diameter growth (apart from the two stringybarks), straightest stems, least malformation, good crown health and largest number of potential sawlogs per tree of all groups. The peppermints were generally slower-growing and more sinuous than the ashes. Some of the gums grew well and most survived better on this adverse, eroded site than the other groups. Superiority of *E. fastigata* and *E. obliqua* was confirmed by other trials in Hawke's Bay and the Wairarapa region. This was an interesting trial, outside normal forestry country at that time and our previous experience.

Heather McKenzie originated a sawing study at Golden Downs (New Zealand northern SI) on *Eucalyptus nitens*. We were interested in the problems and possibilities of growing *E. nitens* for sawn timber and also veneer. The trees had been pruned up to height 8 m and grown for 15 years at low stocking to 57 cm diameter at breast height. This stand provided 15 trees, preselected for a range of wood density. Lumber and sliced veneer were cut from the 5 m butt logs, veneer was rotary-peeled from the second logs from height 7 to 13 m, and each tree was evaluated for production of appearance-grade lumber and veneer.

Butt-log quality was good, as pruning had effectively restricted the knotty core, and there was little decay from branches in either butt logs or veneer billets. Longitudinal growth stresses varied widely among trees, resulting in serious log end-splitting and sawlog flitch movement during sawing (spring), which led to crook in sawn timber, substantially reducing timber conversion in some trees. Collapse and internal checking were prevalent in air-dried lumber, and numbers of checks varied widely among trees. Face-checking was found in boards from all trees after kiln-drying and reconditioning, and even those with very few face checks had internal checks.

A 1 m billet from between butt and second logs and discs at successive heights plus breast-height increment cores and longitudinal growth strain served to characterise the wood properties and processing, product and clearwood mechanical properties of each tree. Fibre dimensions, density and microfibril angle were measured by SilviScan at height 6 m.

Log-end splits, flitch movement off the saw, timber crook, and timber conversion percentage showed strong intercorrelations with one another and with (standing tree) longitudinal growth strain. Amounts of internal checking and collapse in the air- and then kiln-dried butt-log boards were strongly correlated with checking, measured on discs and on the billet boards.

Clearwood modulus of elasticity strongly increased from pith to bark, density had a small increase, and microfibril angle had a rapid decrease from the pith over the first seven rings. Modulus of elasticity was strongly correlated with the SilviScan density/microfibril angle ratio.

Luis Gea, Ruth McConnochie and S. Wynyard in 2007 between them opened up a new open-pollinated breeding strategy for *Eucalyptus nitens* in New Zealand. This utilised microsatellite markers to reveal the parental identity of OP forwards selections. Microsatellites are the preferred markers to reveal genetic relationships between individuals, largely owing to their co-dominant inheritance. Forwards selection of individuals for the breeding population and future deployment was simulated using ten open-pollinated seedling offspring from each of ten clones in a clonal seed orchard. A set of 15 microsatellite markers was chosen from the 41 initially tested. Ninety of the hundred progeny sampled matched consistently to a single mother and father but 13 of these were evidently selfs. There was a broad level of outcrossing at the individual and provenance levels, and there appears to be little indication that individual clones favour specific pollens. Estimates of the coefficient of inbreeding and coefficient of co-ancestry were derived for the seed orchard and forward selections.

A sawing study was carried out by Ruth McConnochie and Charlie Low on unreplicated 25-year-old *Eucalyptus fastigata*, *E. globoidea*, *E. muelleriana* and *E. pilularis* from Rotoehu Forest, to determine if these species could be used to produce high-quality timber on shorter rotations (the study was eventually written up by Trevor Jones). The butt and second logs of 15 trees of each species were quarter-sawn and flat-sawn, respectively, and the boards assessed for shrinkage and distortion, visual and mechanical properties and surface hardness.

Growth-stress release during sawing, in *E. fastigata*, combined with end-checking during drying, resulted in board end-splitting that reduced the sawn recovery compared with the other species. This also produced high levels of crook in the quarter-sawn boards of all species. There was no surface checking and little or no drying collapse and internal checking in any species. The proportions of visual clears and No.1 cuttings grades were low, particularly for *E. muelleriana* and *E. pilularis*, due to the presence of knots. The boards of all species had high density, modulus of elasticity and surface hardness and machine stress grades of MSG10 to MSG15. These species have the potential to produce high-quality timber on 25-year rotations, but pruning will be required to improve visual grades so that a higher proportion of boards can be used in appearance applications.

Some 40 years have elapsed since the beginnings of eucalypt tree improvement by Mike Wilcox in 1975. The process has involved a search for 'good' species, good for producing short-fibred pulp and good for producing appearance grade timber. Although *E. nitens* (in Southland) and *E. fastigata* and *E. regnans* in the NI are used for pulp, neither species can compete with *E. globulus* for growth rate and pulp quality, as grown in Australia. For sawn timber, there has been little commercial interest in growing eucalypts, though small growers favour stringybarks like *E. muelleriana* and *E. globoidea.*

As subjects for breeding, the species are more difficult to establish than radiata and suffer many fungal diseases and insect pests. None of the species grown are easy to vegetatively propagate, and their open-pollinated progeny do not follow the pine genetic model, especially in early generations. There is no doubt that eucalypt species need a full rotation of testing before adoption. Many eucalypt species have failed for one reason or another, and there has been very limited success in eucalypt breeding in NZ.

# Chapter 10
# Cupressus and Other Conifers

Martin Bannister established a progeny experiment in Rotoehu forest in about 1960 of a variety of progenies from different parts of the species range in Mexico and Guatemala. He reported initial differences among seedlots in 2009.

Twenty-one-year-old *C. lusitanica, C. macrocarpa* and the Leyland hybrid, *Chamaecyparis nootkatensis* x *C. macrocarpa*, were sawn with some interesting results, particularly that *C. lusitanica* was much less stiff (lower MOE (modulus of elasticity)) than the other species.

There were two sets of 76 OP families forming the base of *C. macrocarpa* breeding, 76 NZ families collected from select trees in NZ and another group of 76 families collected from California. This study looked at genetic parameters of these populations growing at Strathallan in Southland, the first estimates of these parameters in *C. macrocarpa.*

**20. A.J. Carruthers 1964: Establishment plan for a provenance trial of *Abies grandis*.**

Nine provenances were planted on 23 sites between 1958 and 1962. The whole latitudinal range of *Abies grandis* was covered, from California coast, Peter's Creek and Dehaven, 39° 19′ and 39° 39′, to Courtenay, Vancouver Is. 49°, all on coastal sites. Altitudes varied from 400 to 2000 feet. Stock was raised as 2+2 or 2+1 at various nurseries. Nursery heights prior to planting varied from 15 4 in. at FRI for seedlot 457 to 6.5 in. for 723, due to various ages of stock.

One, sometimes two plots were planted of some of the seedlots at a variety of sites. The most complete sites were Whangapoua, Kaingaroa Cpt.1149, Patanamu, Gwavas, Hanmer, Berwick and Rankleburn. At the other sites, 25 provenance plots were planted. Number of trees per plot varied from 100 to 144, occasionally up to 180.

Survivals 1 year after planting ranged from 65% at Gwavas to high 90s at most sites. Survival of overall provenances varied from 72% for 725 to <86% for the rest. The species generally failed.



C. J. A. Shelbourne, M. Carson, *Tree Breeding and Genetics in New Zealand*,
https://doi.org/10.1007/978-3-030-18460-5_10

**L.E. Fung 1993: Wood properties of New Zealand-grown *Cunninghamia lanceolata*. NZJFS 23.**

*Cunninghamia lanceolata* (Chinese fir) is considered one of the most important trees in central-southern China. In China it has been cultivated as a timber species for over 1000 years. The species does not appear to have been planted much outside China and Taiwan. Physical, mechanical, and drying properties of three stands of New Zealand-grown *C. lanceolata* were assessed, and anatomical and pulping studies were reviewed.

Results clearly showed that mechanical and physical wood properties of New Zealand-grown *C. lanceolata* are numerically lower than those of native-grown (Chinese/Taiwanese) *C. lanceolata*. The main factor is a lower basic density and reduced strength. Shrinkage, however, appears to be fairly constant. In comparison to *Pinus radiata*, *C. lanceolata* wood is of lower density and is not as strong.

Shrinkage is similar for both species and there is little degrade under a conventional kiln schedule or air drying. Drying rates are similar to *P. radiata*. It appears that with its low basic density *C. lanceolata* would be unsuitable for heavy structural uses. However, its dimensional stability, ease of drying and reputed durability would allow it to be used in applications such as weatherboarding, panelling and joinery. Its growth rate, however, is low and it is probably poorly adapted to NZ conditions.

***Comment*** An interesting paper from Lindsay Fung, following his degree in China. I know of only one plot in the Long Mile, of *C. lanceolata*. It is such a great species in China (see my tour notes from 1998 and paper on clonal forestry in NZJFS by Chou), but it seems ill-adapted to the NZ climate.

**Bannister, M.H. 2009: Variation in seedlings of *Cupressus lusitanica*. NZJFS 39: 57–64.**

In 1960, seedlings from 28 lots of *Cupressus lusitanica* seed of indigenous provenances in Mexico and Guatemala, and of exotic provenances from Portugal, Kenya and New Zealand, were arranged in a replicated design. At 1 year of age, a visual appraisal led to a tentative classification, the indigenous material being divided into four groups:

1. *C. lusitanica* var. *lusitanica*, from a central region in Mexico
2. *C. lusitanica* var. *benthamii*, from Hidalgo province in Mexico
3. *C. lusitanica* of uncertain status, from Guatemala
4. Two seedlots of doubtful status, believed to be *C. arizonica* from Durango, Mexico

Statistical analysis suggests that in orchard seedlings, number of cotyledons, from two to six per seedling, varied significantly ($p < 0.05$) from one seedlot to another and possibly varied on a regional basis in the wild. Measurements of height, leaning of the stem, length of longest lateral and number of laterals, taken as single variates, all showed significant seedlot variation within and between geographical groups.

Trait means of the 26 seedlots (excluding the two from Durango) taken in pairs, showed significant positive correlations from 0.37 to 0.88. Many of the differences between seedlots of indigenous origin may be ascribed to differences in the intensity of inbreeding. This speculation also applies to the apparently wider variation of the seedlots from cultivated trees. Another possible source of genetic variation in cultivated material is hybridisation between *C. lusitanica* and *C. macrocarpa*, which is known to occur.

***Comment*** This is an interesting bit of history of the Bannister Rotoehu *Cupressus lusitanica* trial, in the nursery in 1960. Seed was collected by various eminent people, including Hugo Hinds, Egon Larsen, Sir Harry Champion, H.H.C. Pudden in Kenya and Jap van Dorsser in the FRI nursery. Thanks to Tom Ransfield for photos and RDB for statistical analysis.

**C.B. Low, H.M. McKenzie, C.J.A. Shelbourne and L.D. Gea 2005: Sawn-timber and wood properties of 21-year-old *Cupressus lusitanica, C. macrocarpa* and *Chamaecyparis nootkatensis x C. macrocarpa* hybrids. Part 1. Sawn timber performance. NZJFS 35.**

Demonstration plots of *Cupressus lusitanica* Mill., *Cupressus macrocarpa* and *Chamaecyparis nootkatensis* Spach x C. macrocarpa in Rotorua, aged 21 years, were felled to compare lumber performance for appearance and structural uses. The trees had been planted at 1111 stems/ha and later pruned in stages to height 5–8 m and thinned to 550 stems/ha.

Twenty trees of *C. lusitanica*, 7 of *C. macrocarpa*, and 12 of Leyland were cut into 3 m sawlogs and sawn to 150 × 50 mm and 100 × 50 mm sizes, slowly air-dried, then kiln-dried and dressed. Lumber was graded visually as appearance and structural grades. All boards were tested for long-span bending stiffness using the E-grader, and a sample was tested for characteristic bending stiffness and strength.

Each taxon had some advantages and disadvantages in growth, form and sawn timber characteristics. *C macrocarpa* had grown to the same diameter at breast height (DBH) as *C. lusitanica*, and both had grown much faster than Leyland. *C. macrocarpa* was the tallest but was badly affected by canker. Leyland had straighter stems than the others, and a higher frequency of branching.

Sawn-timber recovery was 50–60% for all log height classes of each species, except for the butt logs of *C. macrocarpa* where it was approximately 40% owing to fluting and high taper. Leyland yielded more of the best appearance grades, with 46% dressing, 35% merchantable and only 19% Box. *C. lusitanica* averaged 26% Box, and *C. macrocarpa* 46%. Checks within knots were the worst defect for appearance grades in *C. lusitanica*, a surface checks in *C. macrocarpa* and pruned branch stub holes in Leyland. Long-span bending tests showed that *C. lusitanica* boards were much less stiff than those of the other species/hybrids. Bending stiffness of *C. lusitanica* was 4–6 GPa for both board sizes and 6–8 GPa for *C. macrocarpa* and Leyland.

Stiffness increased from the inner boards to the outer in *C. lusitanica* (4.3–7.2 GPa). Characteristic bending strength was lowest for *C. lusitanica* (21.3 MPa) and values for *C. macrocarpa* (31.4 MPa) and Leyland (28.0 MPa) were similar to global *Pinus radiata* D. Don values.

**L.D. Gea and C.B. Low 1997: Genetic parameters for growth, form and canker resistance of *Cupressus macrocarpa* in New Zealand. NZJFS (27).**
In two *C. macrocarpa* Gordon progeny tests planted at Strathallan (South Island) and Gwavas (North Island), with 76 New Zealand land race families and 76 Californian families, the land race progenies performed better than the Californian progenies. Among Californian populations Point Lobos showed flatter branch angles than Cypress Point.

Most traits were highly variable and the narrow-sense heritabilities for branching (0.5) and straightness (0.3) were high to moderate. Due to micro-site variability, diameter growth and height showed smaller heritabilities. As can be seen from the internal reports, there was a considerable amount of work done on clonal testing of *C. lusitanica.*

## References

20. A.J. Carruthers 1964: Establishment plan for a provenance trial of *Abies grandis*.
L.E. Fung 1993: Wood properties of New Zealand-grown *Cunninghamia lanceolata*. NZJFS 23.
Bannister, M.H. 2009: Variation in seedlings of *Cupressus lusitanica*. NZJFS 39: 57–64.
C.B. Low, H.M. McKenzie, C.J.A. Shelbourne and L.D. Gea 2005: Sawn-timber and wood properties of 21-year-old *Cupressus lusitanica, C. macrocarpa* and *Chamaecyparis nootkatensis x C. macrocarpa* hybrids. Part 1. Sawn timber performance. NZJFS 35.
39587 C.B. Low and S.J. Gatenby 2006: Assessment results of the 1985 cypress species trial at 3 locations.
L.D. Gea and C.B. Low 1997: Genetic parameters for growth, form and canker resistance of *Cupressus macrocarpa* in New Zealand. NZJFS (27).

## Internal report references, not available as PDFs

1291 D.C. Maika 1999: Trial establishment of 2nd. generation progeny test of *C. lusitanica.*
30806 J.T. Miller 2000: Cypress strategy.
30424 C.B. Low and D.C. Maika 2000: Trial establishment report of 2nd. generation progeny test of *C. macrocarpa*
32928 B.V. Geard 2001: Comparison of the growth rate, form, branching, health and wood characteristics of clones and seedlings of *Cupressus spp.*
35199 G.T. Stovold and D.G. Holden 2002: *C. lusitanica* clonal trial establishment.
36398 G.T. Stovold and D.G. Holden 2004a: Establishment of *C. lusitanica* clonal trial.
36406 G.T. Stovold and D.G. Holden 2004b: Establishment of *C. macrocarpa* clonal trial 2003.
39587 C.B. Low and S.J. Gatenby 2005: Assessment results of the 1985 cypress trial at 3 locations.
37654 T.R. Chandrasekhar, S. Kumar and C.B. Low 2006: Preliminary estimates of genetic parameters of early growth in a clones-in-families test of *C. lusitanica.*
39583 C.B. Low 2006: Age 18 assessment of cypress hybrid clonal trial at Gwavas.
41081 K.R. Fleet, G.T. Stovold and C.B. Low 2007: Establishment report for 2006 *C. lusitanica* progeny trials.
45320 C.B. Low, M.A. Miller and D. Grogan 2009: Establishment report for cypress hybrid trials.

# Chapter 11
# Clonal Forestry

Much research on clonal effects has been done since 1969, which is too much to summarise here. The cloning of progeny test seedlings was simulated to form a cloned breeding population allowing accurate within-family selection and was reported by Shelbourne (1992) and later by Shelbourne et al. (2007).

**R. 374. I.J. Thulin 1969: Breeding *Pinus radiata* through seed improvement and clonal afforestation. Second World Consultation on Forest Tree Breeding, Washington, 7–12 Aug. 1969.R.**
The introduction focuses on poor stem quality, strong basket whorls, steep branch angles and crooked internodes of radiata pine which usually required the planting of large numbers of seedlings to get sufficient well-formed trees for a final crop.

Ib Thulin dates the start of radiata breeding to 1953, with intensive plus-tree selection and clonal seed orchards, and forecasts 10,000 lb of seed by 1976. The assessment of two 11-year-old CP tests revealed improvement of 63% in stem straightness, 45% in branching characteristics and 14% in volume.

He describes cutting propagation with '80% rooting of all clones up to 12 years old'. And 'plantable rooted cuttings from trees up to 15–17 years old can be produced in commercial quantities at a cost of $20/1000 vs. $7/1000 for seedlings'. The oldest cutting propagation trials were then 2 years from planting. He saw maturation giving the bonus of good form and few defects.

He described the '268' selection programme from his view of selecting large numbers of plus trees, aged 12–17 years, to be propagated by cuttings, producing large numbers of cuttings for bulk repropagation. This was written in December 1968 when the '268' cuttings had only been set a few months previously. He believed that there would be 80% strike. He tabulated how 500 clones could be bulked up, with parallel clonal tests reducing the number of clones being propagated and planted commercially. In retrospect, it was all dreadful 'pie in the sky'.

He used the predicted percentage gains from tree breeding methods (an abbreviated version was presented by Ib at this consultation) to reason that clonal forestry through clonal testing and afforestation would provide greater improvement in one



C. J. A. Shelbourne, M. Carson, *Tree Breeding and Genetics in New Zealand*,
https://doi.org/10.1007/978-3-030-18460-5_11

cycle of selection than recurrent selection and clonal seed orchards. He didn't recognise the effects of ageing or maturation. Clonal forestry *(he thought)* has the advantages of a 'quick, effective and relatively cheap method of improvement', good for low $h^2$ traits, disease resistance, wood quality and adaptability to cold and poor nutrition sites.

***Comment*** As I remember things, the cutting propagation of the 12–17-year-old '268' ortets was experimental (their age was against success), and by winter of 1969 it was clear that only a fraction of the 600 clones would produce enough cuttings to plant a small clonal trial, which was done at Cpt. 1350. Their grafting was done so that there would be scion material available for seed orchard grafting and for archives for control pollination, also at Cpt. 1350 Kaingaroa Forest. Later on, the cuttings in the clonal test showed strong effects of maturation.

The '266' clones were planted in 1968 at Cpt. 1350, and growth of the 1+1 cuttings from 5 to 6-year-old ortets had been good, so this probably added to Ib's overconfidence.

I believe we quite quickly realised that an extensive clonal forestry programme as described wasn't going to happen, but we continued to try with selections from such series as the '870', '871' and '873' propagated from CP seedlings of 4, 5 and 7 years from seed and subsequently immediately hedged. These clones were eventually planted, successfully from hedges, as clonal tests at Cpts. 1350 and 327 but could not be repropagated without maturation problems.

**C.J.A. Shelbourne 1969: Tree breeding methods. Technical Paper 55, Forest Research Institute, New Zealand Forest Service, Wellington.**
This was a 'Review of tree breeding strategies in relation to classical plant breeding methods, with quantitative genetic expectations of gain...' (Shelbourne 1969) but is quoted here particularly for the gains predicted for clonal forestry.

Allard (1960) enumerates plant breeding methods with cross-pollinating species which included mass vegetative propagation and with clonal test selection (which applies to clonal forestry). Genetic gain equations are given for each of these methods:

Selection and mass vegetative propagation (clonal selection using whole additive and non-additive variance) and then mass propagation of these clones without further clonal testing.

Ditto plus vegetative propagation of clonally-tested best clones, i.e., clonal forestry.

Gain expectations are shown for #7 and #8 and examples given in the radiata pine examples (Shelbourne, 1969).

***Comment*** Clonal forestry requires some means of storage of clones during the long period of clonal testing, something that eluded radiata researchers until embryogenesis became operational and cryogenic storage in liquid nitrogen was feasible.

**R. 766. C.J.A. Shelbourne and I.J. Thulin 1974: Early results from a clonal selection and testing programme with radiata pine. N.Z. J. For. Sci. 4 (2); 387–398.**

216 trees were selected in 1965 for wood density from a total of 500. These clones, designated as the '266' series, were selected in unimproved 5- and 6-year-old stands and 6–7 years from seed. All clones were planted in two field trials in June 1968 in a single-tree layout, with 2–5 plantable rooted cuttings per block (and two seedling control lots) in 3 blocks per site (Cpt. 1350 and Whaka forest) (Shelbourne and Thulin 1974).

Broad sense $H^2$ were height age 4 from planting 0.40, DBH 0.35, straightness 0.26 and branch cluster number 0.41. Repeatability of clone means for different traits varied from 0.71 to 0.86, based on 9 ramets/clone. Mean height of 1/0 seedlings was less than the clones at time of planting, but by age 4 years, the seedlings were growing faster in height than the clones.

These '266' clones were repropagated 3 years after initial planting, but only 51 clones out of 216 were plantable in a new trial at Cpt. 905 Kaingaroa with the rest in hedged archives. Repropagation of these clones from hedges did not give repeatable results, and there was advanced maturation occurring in the repropagated clones. Although this was an unsuccessful attempt at clonal forestry, this trial, after getting normal pruning and thinning, by age 26 years was valuable material for a clonal sawing study.

> In an applied selection programme, it is essential that select clones retain their superiority through repeated propagations otherwise the viability of the whole scheme becomes questionable. (*I.J Thulin*)

**I.J. Thulin, A. Firth, T.G. Vincent and C.J.A. Shelbourne 1973–1977: Cutting propagation of clones selected in 1968-planted full-sib block plantings at Cpt. 1350, Kaingaroa forest with subsequent establishment of hedges.**

I can't find reports of this, but it was an important attempt at clonal forestry which generated large successful clonal tests at Cpt. 327 and Cpt. 1350 Kaingaroa\. Although the hedging was initially successful in producing cuttings for these clonal tests, with some ageing effects, it was not effective for long-term repropagation.

The first propagation was from young ortets, where they were growing in 100 tree family blocks in Cpt. 1350, selected at ages from seed of 4 ('870'), 5 ('871') and 7 years ('873'). The clones were kept hedged, and clonal tests of cuttings from the hedges, with over 100 clones each, were planted in 1976 at Cpts. 327 and 1350 Kaingaroa. A selection of 10 clones from the clonal test in Cpt. 327 in a STP design was the basis for large mechanical pulping and kraft pulping studies when the trees were 16 years old (Thulin and Firth et al. 1973-77).

**R.D. Burdon and J.M. Harris 1972: Wood density of radiata pine clones on four different sites. NZJFS 3 (1972).**

Wood density of 18 clones was studied on four contrasting sites. Clonal repeatabilities were all 0.75 or greater. Site effects were large, but clone x site effects were minor. Density was little affected by differences in ring width (Burdon & Harris 1972).

**R. 768. R.D. Burdon and C.J.A. Shelbourne 1974: The use of vegetative propagules for obtaining genetic information. NZJFS 2: 418–425.**
In theory, the complete lack of genetic variation among the ortet and the ramets of a clone, in contrast to the considerable genetic variation among seedlings in a family, makes for greater precision and efficiency in obtaining genetic information. On the other hand, vegetative propagation often confounds various non-genetic effects, which are also clonal, with the purely genetic differences.

Two types of information include genetic parameters (variances and covariances among traits) and genotypic values of individuals.

The model:

C = G + M + GM where M is maternal effect common to all ramets of a clone, e.g. ortet age.
Phenotype: P = C + m + E + Cm + CE + mE where m is maternal effect peculiar to individual propagule.

Variances are the same. We then went on to apply the model to a half-sib progeny trial and a clonal trial with appropriate breakdown of the additive and non-additive variances. The outcome is a ratio of between-clone to within-clone mean squares which is much greater than the ratio of between-progeny to within-progeny mean squares. Put differently, clones need far fewer ramets per clone than seedlings per progeny to characterise clones or progenies. With low $h^2$, a clonal trial will be more efficient than a progeny test, provided there is no topophysis. A full-sib trial will be more efficient than half-sib but will inflate between-progeny variance if non-additive variance is important (Burdon and Shelbourne 1974).

We applied the developed model to comparing progeny tests with clonal tests for estimating parent tree genotypic values and noted that single full-sib families give no valid estimates of genotypic values of parents.

**M.J. Carson 1986: Advantages of clonal forestry for *Pinus radiata*. Real or imagined? NZJFS 6 (1986).**
The potential advantages of using tested clones, compared with existing OP and CP orchard strategies, were considerable; clonal forestry shares with CP orchards the advantages of shorter plant production times, control of pedigree and high multiplication of valuable crosses. However, a CP orchard combined with vegetative multiplication is currently proving more cost-efficient than potential clonal forestry in establishing managed stands and technologically is much easier to implement. It multiplies full-sib families rather than the genotypes of CF (clonal forestry), and the clones of CF require an additional cycle of clonal testing (Carson 1986).

Most of Libby and Rauter's list of 18 advantages are easy to negate in the light of then current radiata plant production methods. Clonal forestry shares with CP orchard strategies advantages of shorter plant production times, (*comment.* Not sure about this? Surely CF requires a whole cycle of clonal testing and selection on top of GCA test of parents) control of pedigree, flexibility of deployment, multivalued crosses and efficient capture of additive gains. CF may have additional advantages in increasing uniformity, allowing clone/site matching, controlling growth habit and

**Fig. 11.1** Two clones in FRI Long Mile area, same age, Cyclaneusma-affected on right

capturing non-additive genetic gain. However, a CP orchard strategy is more cost-efficient. The use of clonal forestry requires evidence of greater economic gains than from CP orchards (Fig. 11.1).

**R.D. Burdon 1988: When is cloning on an operational scale appropriate? Proc. IUFRO Conf Breeding tropical trees. Pattaya, Thailand 28 Nov. - 3 Dec. 1988.**

***Comment***: This was a good examination of the use of cloning in vegetative multiplication (VM (vegetative multiplication) of unidentified genotypes) and clonal forestry, CF. He takes a hard look at the requirements of CF with its various problems as a component of a breeding strategy and contrasts those species where vegetative propagation and the problems of maturation are easier to handle than seed propagation. He looks at potential advantages and disadvantages of use of clones in limited numbers and spells out the algebra of clonal selection for CF (and additionally the whole QG (quantitative genetic) picture of gains from selection among and within HS and FS families).

**R. 2402. C.J.A. Shelbourne 1992: Genetic gains from different kinds of breeding population and seed or plant production population. South African Forestry Journal 160 March 1992.**

This was a paper about deterministic gain prediction, using the same equations of Namkoong et al. (1966) and more. The scenarios were of different OP and CP breeding populations, seedling and cloned. Gains were also predicted for derivative clonal and seedling orchards, and in this context, for clonal forestry. These were done under contrasting heritabilities of 0.1, 0.2 and 0.4.

If the BP families were cloned, with ten clones per family with ten ramets per clone, the predicted gain (in the $h^2$ 0.2 selection) increased from about 8% to 14.1%. Only by cloning could one get some effective selection done within family, to boost BP gain.

In the production population options, a clonal orchard from intensive among-family and within-family selection gave gains of 15.5% ($h^2 = 0.2$). Clonal selection in a clonal forestry scenario from OP families in the original OP test increased the gains dramatically above those from seed orchards, gains depending a lot on how many ramets per clone were used. However, these would require ongoing clonal testing and then repropagation of best clones, taking another test cycle, equivalent to backwards selection for a 1.5 generation seed orchard. Gain prediction for clonal forestry does bring out the importance of cycle length (Shelbourne 1992).

**R.D. Burdon, R.D. Gaskin, C.B. Low and G. Zabkiewicz 1992: Clonal repeatability of monoterpene composition of cortical oleoresin of *Pinus radiata*. NZJFS 22 (1992).**

Forty-five clones were studied for $H^2$ of levels of individual monoterpenes using hedged cuttings from 7- to 9-year-old ortets. $H^2$ levels were all very high >0.85. Ramet/ortet repeatabilities were also very high for β-pinene, carene and limonene, lower for α-pinene (Burdon et al. 1992).

**R. Beauregard, R. Gazo, M.O. Kimberley, J. Turner, S. Mitchell and C.J.A. Shelbourne 1999: Clonal variation in the quality of radiata pine random width boards. Wood and Fibre Science, 31(3) 1999.**

Ten clones and two ramets per clone ('266' series whose clonal test was planted in 1968) were felled and sawn at age 27 years, and broad sense heritabilities of board quality traits were documented. Knot frequencies were far higher from trees with shorter internodes and internode length appeared highly heritable. For all log types, the best performing clone for random-width board quality was one with large DBH with the longest internodes, while the small DBH clone with the shortest internodes was the worst. Defect frequency and grade in appearance lumber show high broad sense heritabilities. Value can be predicted from tree internode index which has been shown to respond well (in a 1970 selection programme) to selection and breeding (Beauregard et al. 1999).

**C.J.A. Shelbourne 1997: Genetics of adding value to the end-products of radiata pine. Proc. IUFRO Genetics of Radiata Pine, Rotorua, New Zealand, 1–4 December 1997. (FRI Bulletin No. 203).**

This paper summarises broad sense heritabilities and predicted gains of over 90 different traits from many different studies involving two sets of clones. These were propagated in attempts to institute clonal forestry by GTI from 1966 to 1973. Each set comprised ten clones, two or four trees (ramets) per clone. The first set was felled at age 16 years ('870', '871', '873' series, propagated initially at age 4 years from seed) and was used for a large number of wood property and pulping studies.

The second set ('266' series) was of ten clones of two ramets per clone, originally propagated from 6-year-old ortets and aged 27 years when felled for a sawing study.

Broad sense heritabilities and predicted gains from clone mean selection for clonal forestry are given for 90 different traits, including characteristics of tree growth and morphology, wood and chemical properties and kraft and thermo-mechanical pulping properties (all from 16-year-old ramets of selected ten clones). The sawing study of random-width sawn boards was from 27-year-old trees of the '266' series. The clones were chosen from each series to give a wide coverage of trait variability in the studied trees.

Broad sense heritabilities were generally high to very high, and heritabilities of clone means based on five ramets per clone were mostly over 0.8. The parameter that limits gain from selection is generally the coefficient of variation of that trait.

This paper is important to potential users of clonal forestry deployment, as the 90 or so characters covered are mostly for end products. This collection of trait heritabilities is also valuable for tree breeders generally as, in the absence of non-additive genetic variance of most traits, the ranking of broad sense heritabilities will parallel that of narrow sense heritabilities. With some exceptions, most of the utilisation traits have broad sense heritabilities exceeding 0.6 (Shelbourne 1997).

**B.S. Baltunis, H. Wu, H.S. Dungey, T.J. Mullin and J.T. Brawner 2009: Comparisons of genetic parameter and clonal value predictions from clonal trials and seedling base population trials of radiata pine. Tree Genetics and Genomes. Jan. 5: 269–278.**

DBH was assessed at age 5 years for clones and mostly at age 8 or older for seedlings. Significant additive, dominance and epistatic effects were estimated for DBH. Non-additive effects for DBH were 87% as large as additive effects. Narrow sense $h^2$ and broad sense $H^2$ were 0.14 and 0.26, respectively. Accuracy of predicted clone values was increased by combining clone and seedling data. Best 5% of clones gave gains of 24%, double that of selection on family alone (Baltunis et al. 2009).

***Comment*** One or two queries: Dean et al. (2006) found increased additive relative to dominance variance with age. Sue Carson (1991) found non-additive genetic variance decreasing to near zero in 8-year or older assessments of '850' diallels over ten sites. Wilcox et al. had a similar experience much earlier. Similar experience was recorded with the disconnected diallels (with about 100 parents) among the '875's by Mike Carson (see in Radiata Breeding section). Evidence from genetic parameters for DBH has pointed to marked reduction in non-additive with age. There seems little chance that non-additive effects of 87% are correctly estimated.

Not a lot of information was given about the planting stock used beyond. 'In the current study all attempts to reduce impacts of "c" effects, including taking cuttings from hedges' (Baltunis, Wu, Dungey et al.). The experience with taking cuttings from hedges was more maturation in cuttings taken from a hedge. I believe that most of the Baltunis clonal tests are seriously 'aged', which would mean that the trait 'DBH' is something different from that in the seedlings. I do remember that we got a lot of 'extra' within-family genetic variance in some clonal tests that couldn't be explained by ordinary QG models. Repropagating clones from hedges maintained for this purpose after some years gave poor results. There is a lot of anecdotal information available about these early attempts at clonal forestry from 1966 to 1980.

**Two recent exchanges between Tony Shelbourne and Mike Carson:**

*Tony*

I had some difficulty in understanding parts of the Baltunis et al. ‘Comparisons of genetic parameters and clonal value prediction….’

I didn’t really understand how they combined clonal and seedling parameters from different materials and trials. I do remember that, with the ‘850’ disconnected diallels, Sue (Carson) found minimal non-additive variance at about age 8+ for DBH and the same was true in the ‘875’ diallels. (Didn’t you analyse this material?) Even Mike W. in the little 4 × 4 factorial in Cpt. 1350, planted 1968, found non-additive diminishing rapidly with age.

The second worry I have is that maturation will have advanced considerably in cuttings from nursery stools. Won’t this maturation have affected parameter comparisons with seedlings? I also think that Baltunis estimating non-additive for DBH at age 5 will bias the estimate of non-additive upwards compared with seedlings at age 8+ (his ‘87% greater non-additive versus additive’?)

*Mike*

You’ve raised some really good questions. My comments on your comments:

How did Baltunis et al. combine the data? I’m sure they used BLUP (maximum likelihood), which does have the nice feature of handling imbalance in sample sizes and structures – but also can be a ‘black box’ that can risk the ‘garbage in-garbage out’ outcome. However, in this case, their data were probably good enough to give them genetic parameter estimates that would have been estimated pretty similarly using the older least squares and other methods that you and I are more used to. Yes, Sue estimated SCA/non-additive effects in the 850 diallel at age 8 as being much less than the GCA effects, and I got a similar result with the 875s. I recall that John King confirmed that the SCA (for DBH) in the CNI (Central NI diallel) trials diminished with trial age – which led me to speculate that some of the SCA we thought we were seeing in the trials at a younger age was more likely error.

Rematuration of cuttings from stoolbeds, there are plenty of examples of this having a likely effect on genetic variance. He went on from there on ‘Development of a clonal breeding programme’ (‘266’ clones) with selection of clones on the basis of total genetic variance and repropagation of best clones, gave parameter estimates (for growth mainly) – including the earlier round of GTI clonal trials (which I think you wrote up?). However, with the somatic embryogenesis clones used by Forest Genetics, we were able to keep them relatively juvenile, and with good stoolbed practice (i.e. low hedging and turnover after 3–4 years), we didn’t see much evidence of ageing – and I looked hard for it in the trials, particularly noting that the onset of flowering was similar in the clones to that for the seedling controls. That said, we still seemed to be seeing more additive variance in the clones than we might have expected from progeny trial estimates, and I'm still unsure whether that can be explainable as some form of 'C effect'.

Recent work on other plant species could be indicating that there are epigenetic effects that could be contributing(?). For other traits (i.e. than growth rate), the clonal trials seemed to be giving us quite believable parameter estimates, consistent with those in progeny trials. Notably, neither wood density nor *Dothistroma* resistance showed much/any evidence for non-additive variance.

I do agree with your comment that his age 5 estimate for DBH would likely have been an over-estimate.

**M.J. Carson, S.D. Carson and C. Te Rini 2005: Successful varietal forestry with radiata pine in New Zealand. NZ J. For. May 2005 60–1.**

Varietal forestry (VF) is commercial production and deployment of plants of field-tested individual genotypes of forest tree species. The programme of Forest Genetics (FG) is one of a few successful programmes worldwide, based on 60 years R&D. Clonal variation provides additional benefits over seedlots from OP and CP seed orchards (Carson et al. 2005).

This is a comprehensive paper covering all aspects of the 'varietal forestry' programme of FG. Attributes of VF, especially cryogenically stored clones or emblings, planted in nursery beds to produce cuttings for production are:

Cuttings have some maturation, more erect, less malformation, longer needles, etc.
Customer focus-what the grower wants to produce, disease problems and products.
Risk management; clonal mixtures versus single clone blocks.
Technical, of embryogenesis, to emblings to cuttings.

Benefits and opportunities are:

DBH of 3 clones: 21.6 to 23.6 cm versus OP 20.7 and CP 20.6
Stiffness 7.9 to 8.1 versus 5.7 and 5.9 CP
Density 345 to 373 versus 318 and 324CP
Dothistroma: better 6 clones much improved over OP and CP orchard.

**B.C. Baltunis and J.T. Brawner 2010: Clonal stability in *Pinus radiata* across New Zealand and Australia. 1. Growth and form traits. New Forests DOI 10 1007.**

To investigate GxE interaction for clones of RP, stability was evaluated for growth and form traits in three trials in NZ and three in Australia. 215–245 clones were planted in the 3 NZ trials, 44–69 clones were from 7 families in total in Australian trials, 17–26 clones were in common in 3 Australian (Aus.) trials, 32–52 clones were in common to NZ and at least 1 was in Australian trial. Clonal repeatability ($H^2$) was highest for height (HGT) followed by DBH, STR and BRQ. There was little evidence of GxE for DBH and HGT, most for form traits (Baltunis & Brawner 2010).

Type B correlations, for DBH average, were 0.82 and 0.76 for HGT in NZ.

Within Australia there is only one significant Type B correlation for DBH, and Type Bs were higher for form traits. One only Aus. trial showed high clonal stability with three trials in NZ in HGT. Stable clones can be identified that perform well in Aus. and NZ.

## References

R. 374. I.J. Thulin 1969: Breeding *Pinus radiata* through seed improvement and clonal afforestation. Second World Consultation on Forest Tree Breeding, Washington, 7–12 Aug. 1969.

C.J.A. Shelbourne 1969: Tree breeding methods. Technical Paper 55, Forest Research Institute, New Zealand Forest Service, Wellington.

766. C.J.A. Shelbourne and I.J. Thulin 1974: Early results from a clonal selection and testing programme with radiata pine. N.Z. J. For. Sci. 4 (2); 387–398.

I.J. Thulin, A. Firth, T.G. Vincent and C.J.A. Shelbourne 1973-1977: Cutting propagation of clones selected in 1968-planted full-sib block plantings at Cpt. 1350, Kaingaroa forest with subsequent establishment of hedges.

R.D. Burdon and J.M. Harris 1972: Wood density of radiata pine clones on four different sites. NZJFS 3 (1972).

R. 768. R.D. Burdon and C.J.A. Shelbourne 1974: The use of vegetative propagules for obtaining genetic information. NZJF 4 (2): 418–425.

M.J. Carson 1986: Advantages of clonal forestry for *Pinus radiata*. Real or imagined? NZJFS 6 (1986).

R.D. Burdon 1988: When is cloning on an operational scale appropriate? Proc. IUFRO Conf Breeding tropical trees. Pattaya, Thailand 28 Nov. - 3 Dec. 1988

R. 2416. S.D. Carson 1991: Genotype x environment interaction and optimal number of progeny test sites for improving *Pinus radiata* in New Zealand. NZJFS 21(1):32–49.

R. 2402. C.J.A. Shelbourne 1992: Genetic gains from different kinds of breeding population and seed or plant production population. South African Forestry Journal 160 March 1992.

R.D. Burdon, R.D. Gaskin, C.B. Low and G. Zabkiewicz 1992: Clonal repeatability of monoterpene composition of cortical oleoresin of *Pinus radiata*. NZJFS 22 (1992).

R. Beauregard, R. Gazo, M.O. Kimberley, J. Turner, S. Mitchell and C.J.A. Shelbourne 1999: Clonal variation in the quality of radiata pine random width boards. Wood and Fibre Science, 31(3) 1999.

C.J.A. Shelbourne 1997: Genetics of adding value to the end-products of radiata pine. Proc. IUFRO Genetics of Radiata Pine, Rotorua, New Zealand, 1–4 December 1997. (FRI Bulletin No. 203)

B.S. Baltunis, H. Wu, H.S. Dungey, T.J. Mullin and J.T. Brawner 2009: Comparisons of genetic parameter and clonal value predictions from clonal trials and seedling base population trials of radiata pine. Tree Genetics and Genomes. Jan. 5: 269–278.

M.J. Carson, S.D. Carson and C. Te Rini 2005: Successful varietal forestry with radiata pine in New Zealand. NZ J. For. May 2005 60–1.

C.J.A. Shelbourne, C.B. Low, L.D. Gea, and R.L. Knowles 2007: Achievements in forest tree genetic improvement in Australia and New Zealand: 5: Genetic improvement of Douglas-fir in New Zealand. Aus. For. 70, 1 pp. 28–32.

B.C. Baltunis and J.T. Brawner 2010: Clonal stability in *Pinus radiata* across New Zealand and Australia. 1. Growth and form traits. New Forests DOI 10 1007

R. 1599. R.D. Burdon and G. Namkoong 1983: Short note: Multiple populations and sublines. Silvae Genetica 32(5–6):221–222.

Namkoong, G., Snyder, E. B. and Stonecypher, R. W. 1966. Heretability and gain concepts for evaluating breeding systems such as seedling orchards. Silvae Genetica, 15: 76–84.

R.W. Allard. 1960. Principles of Plant Breeding. John Willey and Sons Inc., New York.

# Chapter 12
# Breeding Theory

Shaw and Hood's (1985) paper vusing stochastic simulation in maximising gain per effort by using clonal replication in genetic tests was an important wake-up call. Cloning in within-family selection is really the only way to do selection without reducing genetic variance in the BP. Shelbourne's (1992) paper was a first attempt at quantifying gains from cloned versus seedling breeding populations. This strategy has been finally adopted for within-family selection in radiata pine. Carson, Vincent and Firth's (1992) paper on control-pollinated and meadow seed orchards of radiata pine was the first to document the new control-pollinated seed orchards which had such major effects on radiata breeding. Rosvall, Lindgren and Mullin's (1998) paper on stochastic simulation of gain in a cloned BP of Norway spruce demonstrated the sustainability, robustness and efficiency of a multigeneration breeding strategy based on within-family clonal selection and as such was a very useful model for radiata pine breeding.

**C.J.A. Shelbourne 1969: Tree breeding methods. Technical Paper 55, Forest Research Institute, New Zealand Forest Service, Wellington.**
This was a review of tree breeding strategies in relation to classical plant breeding methods, with quantitative genetic expectations of gain, as well as predicted gains using data from *Pinus radiata* (Shelbourne 1969).

Allard (1960) enumerates plant breeding methods with cross-pollinating species as:

> Mass selection, Backcross breeding, Hybridisation of Inbred Lines, and Recurrent Selection (RS) as simple RS, RS/GCA, RS/SCA and Reciprocal Recurrent Selection (RRS). Also included are mass vegetative propagation with clonal selection.

Genetic gain equations are given for:

1. Mass selection.
2. Mass selection and progeny testing (an OP seedling orchard using mass selection, OP family selection and within-OP family selection).
3. Simple recurrent selection (a clonal seed orchard without any progeny testing and roguing, seed production area or seed stand) (Fig. 12.1).



C. J. A. Shelbourne, M. Carson, *Tree Breeding and Genetics in New Zealand*,
https://doi.org/10.1007/978-3-030-18460-5_12

**Fig. 12.1** Amberley control-pollinated clonal seed orchard, about 1990

4. RS/GCA. Estimating GCA for backwards selection of orchard clones, i.e. 1.5 generation orchards or roguing first-generation orchard.
5. RS/SCA exploiting dominance variance (not additive only, as in RS/GCA). Two-clone orchard, for exploiting SCA (echoed much later by control-pollinated orchards).
6. Reciprocal recurrent selection (RRS, not really covered).
7. Selection and mass vegetative propagation (clonal selection using the whole additive and non-additive variance) and then mass propagation of these clones without further clonal testing.
8. Ditto plus vegetative propagation of best tested clones, i.e. clonal forestry. The first part of the gain was from phenotypic selection (using all genetic variance, additive, dominance, etc.), and the second part of the gain is from within-family selection, useful in a cloned breeding population.

***Note*** Some reference is made to use of cloning for genetic information.

This paper followed Namkoong, Stonecypher and Snyder (1966) closely with some modifications and additions, and Shelbourne gave the equations for each, taking into account the truncation of additive variance by reselection in backwards selection (in orchard roguing and 1.5 generation orchards).

The terms 'backwards' and 'forwards' selection were not yet in use. The prospective benefits of this work were to get breeders thinking about the role of selection, heritability and genetic gain in planning a programme. Gain predictions for different scenarios were given using radiata genetic parameters for straightness (from Bannister's 14-year-old OP progeny test): $h^2$ of 0.6 for straightness and $h^2$ of 0.19 for DBH and with stated selection ratios.

This paper and the thinking behind it represented a fresh look in 1968 at devising a breeding strategy for radiata. I tried to apply these quantitative genetic principles and gain expectations to planning a programme from scratch.

It was clearly evident that if progeny testing was directed at creating new 1.5 generation orchards, then numbers of genotypes under test needed to be sufficient to get real gains from backwards selection. This led to large numbers of selections and the '268' open-pollinated testing strategy, both somewhat revolutionary at the time. At that time, I don't think we saw this large population of OP families as a 'breeding population'. Rowland was the first to use this term/concept in 1970. This methodology and thinking should have pervaded the new Douglas-fir and *P. contorta* programmes (but didn't) which both got started in 1968 when this paper first emerged as a Branch Report.

Looking back, this paper or, rather, Namkoong et al.'s equations were the start of ongoing 'deterministic simulation' of gains for breeding populations, seed orchards and clonal forestry, as in Shelbourne (1992) for eucalypts at the Durban meeting and for radiata in the 2007 paper in *Silvae Genetica* (about gains from cloning BPs, Elites and orchards).

Deterministic simulation wasn't the most sophisticated simulation tool and only did one generation at a time, in contrast to multigeneration stochastic simulation. It did, however, provide a way of comparing different strategies for BPs and orchards, and for clonal forestry, and I did understand it.

**41. C.J.A. Shelbourne and F.R.M. Cockrem 1969: Progeny and clonal test designs for New Zealand's tree breeding programme.**
This was a justification for open-pollinated progeny testing for GCA estimation and reselection of seed orchard clones versus various forms of control-pollinated progeny tests. The philosophy of testing large numbers of less intensively selected parents is outlined (Shelbourne and Cockrem 1969).

The adoption of the Blocks (Sets) in Replications design of Schutz and Cockerham was proposed as a means of reducing block size and number of entries per block, thus reducing environmental variation within a block.

Using ten-tree plots for progeny or three-tree plots for clonal tests is proposed. Balanced or partially balanced incomplete block designs were deemed unsuitable on practical grounds. Design models and expected MS are given.

**R. 772. C.J.A. Shelbourne 1973: Problems and prospects in the improvement of forest tree species. Proc. 2Nd General Congress, SABRAO, New Delhi 1973. reprinted from Indian J. Genet., 34A 1974.**
(*Note*. I think this paper was written in 1972, as the SABRAO Conference was postponed. Burdon and Shelbourne NZJFS (1971) is referred to in a long list (33) of quite topical references, mainly 1966 to 1971.)

This was a 'global' look at the state of genetic improvement of forest trees for the benefit of the plant breeders and other crop research people, at this Society of Applied Breeding in Asia and Oceania meeting in New Delhi in February 1973.

There were very few Indian tree breeders, Kedharnath being the one who invited me. This paper is a useful round-up, in 1972, of a wide variety of issues in tree breeding, which was still a very new discipline/science/ practice, hence its inclusion here. Shelbourne's experience in Zambia with tropical pines and eucalypts and especially the work with the NC State University-Industry Cooperative Tree Improvement Programme, plus the 4–5 years in NZ, was all relevant.

After the introduction, annual plant breeders needed to be put into the tree picture, with trees' big size, long rotation, plantation vs native forest, different products and competition in trials between individuals and plots. Genetics however is in common with many crop plants and most animals, involving a diploid, outcrossing, heterozygous breeding system and the same quantitative genetic theory but still often only the first generation from the wild. The 'problems and prospects' listed and described were:

A. *Selection of species and of provenances, exotics* vs *native spp.*

Our problems were with choice of best provenances in youngish trials vs selecting in maybe suboptimal local land races, as in Douglas-fir and *P. contorta.* (I supported our choice of selecting within local land races, used at the time, though now I have serious doubts about it.)

B. *Choice of selection criteria*

Species-/provenance-level selection criteria may be different from those for within-population selection. There are serious problems of economic weighting and genetic weights in an index. A Stonecypher quote was 'Availability of reasonably accurate economic information (on different traits) could well be the limiting factor in applying multi-trait breeding...'. Heritability information is needed and is only just becoming available from family experiments of increasing age; there were serious problems in getting good estimates. More experiments and more families were needed. Gain predictions were also needed but insufficient genetic parameters had been published.

C. *Initial selection and multigeneration breeding plans*

Most tree breeding programmes had been initiated in the last 20 years, many in the last 10. The old style of intensive plus-tree selection has predominated, resulting in some cases in too few clones for either roguing or new 1.5 generation orchards. Selection methods have not yet been developed for producing offspring for the next generation, and mating designs for estimating GCA are no good for advancing generations.

D. *Improved seed production*

There are several questions, not yet all resolved: clonal vs seedling orchards, seed sands, isolation of orchards, seed tree selection, random male parentage and graft incompatibility.

E. *Family experiments*

Objectives were often not properly defined. These should be to demonstrate gain, estimate GCA of parents, provide trees for advanced generation selection and estimate genetic parameters, and some or all of these. What mating designs should be used?

F. *Cloning*

Libby's (1964 and 1969) papers were an important extension of quantitative theory, including topophysis and use of grafts and cuttings for orchards and for multiplication.

G. *Hybridisation*

The paper did some debunking of interspecific hybrids for tree improvement, though inter-provenance hybrids may be useful for exotics. Inbred lines are generally no use for trees. Conservation of unselected stands of populations under selection is important. The spread of 'dogma' is to be avoided. Conventional wisdoms are difficult to change. Ideas and methods get fixed, especially where early developers had little genetic information. An occupational hazard of tree breeders is the need to make decisions on the basis of slender evidence which may turn out wrong. Breeding programmes must be designed, to be flexible, to allow evolutionary change. The theoreticians can easily become isolated from the practitioners (Shelbourne 1973).

**R. 531. R.D. Burdon and C.J.A. Shelbourne 1971: Breeding populations for recurrent selection: conflicts and possible solutions. NZJFS 1(2):1174–193.**
Numbers of trees initially selected, and progeny tested, strongly influence genetic gains, effective population size, future inbreeding and the size of the next generation's breeding population. In this context, a NZ case study was given, from '850' (1951–1965) to '268' (1968) selections, early and later orchards, with open-pollinated (OP) vs control-pollinated (CP) testing. The need for some sort of CP families for producing second-generation selections underlay their future proposed strategies.

This was a 'seminal' paper, the first to look at a 'breeding population' and what it needed as a population for recurrent selection. The role of 'combined selection' using information from sibs, parents, ramets of ortets and the individual through a selection index (different from a multi-trait index) to get a better estimate of the individual's genotypic value was spelled out, for the first time in the NZ context. We had to discover whether the features of mating designs to provide additional information were worth the 'costs' in the biological constraints of a tree breeding programme. This was put clearly (not for the first time, e.g. animals) but apparently the first time for forest trees.

The curvilinear graph shows selection intensity in selection differential units against number of trees screened per select tree. Rowland developed the concept, considering large effective population size necessary in a situation with many genes (polygenic inheritance) controlling a single trait, and concluded that a BP needed 200 or more genotypes. He compared the various mating designs, open-pollinated (OP), North Carolina 2 (factorial), North Carolina 1 (nested), diallel, partial and disconnected, single-pair mating (SPM), polycross, nested, etc. and compared the SPM with polycross in advantages and disadvantages. He developed four flow-charted strategies using OP, SPM and various types of polycross. All but one were based on use of OP families only to get initial GCA estimates to allow forwards selection in CP families, backwards selection for roguing or establishment of 1.5 generation orchards (he never used OP families themselves to advance the BP to the next generation).

The second flow chart shows a strategy where OP tests estimate GCA of parents and combine these with backwards selection to choose original parents for SPM (single-pair mating) as the main source of trees for advanced genetic selection (AS). He mainly used some form of backwards selection of parents to create the next generation, with OP families being fit (or not) for early GCA evaluation. However, he did use a nested polycross as the basis for immediate forwards selection. He did not distinguish 'backwards' and 'forwards' selection as such and failed to point out the importance of the time factor in advancing the BP and generating GCAs for orchard establishment and roguing. Note: See Burdon and Kumar (2004) for a change of 'tune' about the use of OP for AS.

***Comment*** Libby references to Osborne (1957), plus Libby's assistance during his sabbatical in NZ, helped power this paper. Rowland was first in this area. Mention was made of use of cloning for within-family selection, but this was not considered further, based on Libby's (1969) opinion that more genotypes covered by conventional within-family selection overcame advantages of cloning. However, they didn't look at the time factor and GxE nor simple within-family selection.

**R. 548 C.J.A. Shelbourne 1971: Planning breeding programs for tropical conifers grown as exotics. IUFRO (section 22) Gainesville, Fla. 1971. Symposium on "Selection and breeding to improve some tropical conifers".**
This paper was written in a practical and prescriptive style for breeders of tropical species without a lot of quantitative genetics background. On the other hand, it captured the 'know-how' at the moment of 1970, which had rapidly evolved in NZ about breeding strategy. It draws on the *Tech. Pap. 55 Tree Breeding Methods* itself a development of Namkoong, Stonecypher and Snyder (1966). It is based on the planning of the big '268' open-pollinated selection and testing programme and the adoption of 'sets in replications' design of Schutz and Cockerham.

Burdon's paper presented the theoretical backing of these plans and ideas as well as an NZ case study of the '268' programme. This paper made first use (that I am aware of) of the term 'breeding population' which we had been referring to as 'advanced generation breeding/selection'.

***Comment*** This paper emphasised the importance of choosing the 'right' selection criteria, based on technical requirements of future markets and in the context of a selection index, with economic and genetic bases.

The curvilinear effect on selection differential of proportion saved was spelled out. The various objectives of family experiments were listed, as well as the clarification of 'family plantings for advanced generation selection (AS)' versus 'progeny testing for GCA' (RSGCA). The experience of replanning the NZ radiata programme and initiation of new breeding work in Douglas-fir and *P. contorta* (1967–1969) were all distilled in this 'how-to' paper, in parallel with the Burdon and Shelbourne's paper which properly tackled systematic and theoretical aspects of breeding populations. The importance of these developments, both for radiata and for the nearly simultaneous initiation of breeding in Douglas-fir and *P. contorta*, presented breeding questions for two species with very wide provenance differentiation, in contrast to radiata with the opposite (Shelbourne 1971).

**R. 683 C.J.A. Shelbourne 1972: Genotype-environment interaction: its study and its implications in forest tree improvement. IUFRO Genetics-Sabrao Joint Symposia.Tokyo, October 1972.**

This was another invited conference paper, possibly postponed by a year, about a subject that is important in planning selection of provenances and in tree breeding. About half the 23 A4 pages of text was on the significance of predictable and unpredictable GE, and on methods of measuring and elucidating it, and the other half on published and unpublished results of GE studies of forest trees from a lot of pers. comms. from individual contributors (Shelbourne 1972).

Methods of GE exploration included:

- Ranking entry means
- Correlations between entry means in pairs of environments
- ANOVA components (ratio of GxE/family variance)
- Regression of entry means at different sites on mean of all entries at that site (Finlay & Wilkinson regression, and Eberhart & Russell)
- Comparison of gains within and across environments, implications of species and provenance interaction with site and ditto of family x site interaction

A mass of case histories from tree breeders follows, based on publications and personal communications.

The bottom line seems to be:

- Lots of early GE is detectable (probably in error) in very young trials on too few sites.
- GE in radiata is mostly due to big site fertility differences.
- GE is common in widely distributed native populations of species in PNW USA, especially, involving altitude and continentality.

In breeding work, the problems of GxE are two-fold:

1. How to stratify sites to minimise losses in gain? A one-shot study is needed
2. The other problem recurs each cycle; testing and selection of new genotypes?

Provenance testing should be carried out in developing breeding programmes to identify well-adapted, high-performing groups of populations (especially in an exotic context) from which to breed stable varieties/breeds and balancing costs of stratifying provenances versus not doing so.

A disconnected diallel series of 50 families, planted on 10–12 sites, and the polycross progeny tests of 100 selections on 6 sites should well evaluate GxE effects as well as genetic parameters. The costs of large progeny tests in land, stock, planting and assessment militate against planting operational breeding tests on enough sites to pin down GxE.

**R. 1021. R.D. Burdon, C.J.A. Shelbourne and M.D. Wilcox 1977: Advanced selection strategies. Third World Consultation on Forest Tree Breeding, Canberra and Rotorua.**

This was an invited conference paper which provided a good review of the literature and also gave our ideas of the time on how to advance a breeding population. This was of general application to outcrossing species, and the references came from mostly pine programmes, *P. pinaster* in Bordeaux, *P. taeda* and *P. elliottii* in SE USA, *P. radiata* in NZ and *Picea abies* and *P. sylvestris* in Sweden. Main contributors were Baradat, Libby, Lindgren, Arbez and Mauge, Namkoong, Shelbourne, Burdon, Stonecypher, van Buijtenen and Squillace.

Mating designs for advanced generation selection (AS) include:

- The simple: open-pollinated, polycross, single-pair mating and selfing
- The complex: hierarchical or nested, NC 1 and NC 2 and various diallels, including partial and disconnected diallel

Designs had several purposes, including GCA estimation, genetic parameters and AS. No one design is best for all needs. An account of NZ '268' OP tests was followed by disconnected diallel of 80 '875's (although a true AS experiment, it was not apparently seen in this context). This design was favoured but not without some disagreement over a counter idea of polycrossing followed by SPM. Another good option was disconnected factorials of 4x4. The problems with field designs for AS are that single-tree-plots are optimal for family mean selection, but large plots per family are better for within-family selection. This favours use of complementary experiments. There was some confusion about the different respective roles of GCA ranking and advanced generation crossing; AS assessment and selection are different from GCA ranking of parents. A multi-trait combined index was used for radiata by Wilcox and by Baradat in France and covered by Stonecypher and Arbez in Bordeaux (1976) (all probably involving too much family selection for AS). AS requires intermating of the future parents of the next generation, and parental selection must be low-key or zero.

Sequential culling is still good for expensive traits like density. Early selection is desirable, yet juvenile-mature correlations are often poor, and rapid generation turnover accelerates decline of effective population size. Regarding population size and inbreeding, short-term gain prejudices long-term gain. Reduced population size means more inbreeding eventually. Rowland proposed a hierarchy of populations and creating sublines within the BP (the familiar pyramid). Sublining minimises crossing between relatives in seed production but accepts some inbreeding within sublines. This reduces loss of effective population size in the Main BP.

***Note*** Only within-family selection is the solution for maintaining effective population size.

The main new stuff in 1976 was discussion of mating designs for AS and use of sublines. There was a suggestion of complementary field designs to allow good family mean and within-family selection and combined selection. It was assumed there was a need for an ongoing BP of 200+ parents in good provenances. This points up the features of good AS designs which include single-pair mating. This is really nothing very new but a thoughtful look at advanced selection for BP development which has application to the about-to-be-started Eucalypt programmes by Mike Wilcox (Burdon et al. 1977).

***Comment*** There is nothing here about the question of among- versus within-family selection in a BP, and neither is the possibility of clone-within-family selection mentioned.

**R. 1089. R.D. Burdon 1977: Genetic correlation as a concept for studying genotype-environment interaction in forest tree breeding. Silvae Genetica 26(5–6):145–228.**

'Main attention should, paradoxically, be given to the role of environments rather than genotypes in generating interactions' (Burdon 1977) in contrast to the methods used for crop plants. And he gives the reasons why things are different. He covers traditional ANOVA and then the alternative concept, genetic correlation between environments and its estimation, applications and statistical properties.

Burdon and Low (1973) and Burdon (1975 and 1976) used simple correlation between clone means at pairs of sites to evaluate extent of clone-site interactions. Genetic correlations between traits on the same individual are Type A. Type B correlations are between different individuals of the same genetic group, e.g. family.

Rowland reduced the $R_g$ between sites to Cov g $_{xy,}$ the cross product between environments of family means. This is divided by $h_{gx}$ x $h_{gy}$ which estimates the genetic correlation between those two sites (where $h^2_{gx}$ is the heritability of family means at site x and $h^2_{gy}$, at site y). The genetic correlation between sites can be used for predicting gain from measuring families on one site to gain on another. There are a number of practical benefits of using this concept. It is good for defining breeding regions and predicting gains, and in analysis of data, one can do ANOVA one site at a time and then get genetic correlation between sites, as above.

**R. 1296. R.D. Burdon 1981: Generalisation of multi-trait selection indices using information from several sites. NZJFS 9(2): 145–152.**

This paper is an application of the genetic correlation between sites reported in the SG paper above. A multi-trait index is formulated for selecting parents from half-sib family tests, treating expression of say DBH as a different trait at each site (Burdon 1981). This involved analysing data one site at a time and linking results by $r_g$ between sites. Rowland gives all the algebra supporting this approach and lists the application by Shelbourne and Low (1979) in NZJFS (1981) paper.

**R. 1518. R.D. Burdon 1982: Selection indices using information from multiple sources for the single trait case. Silvae Genetica 31(2–3): 81–85.**

This was a theoretical development of combined half-sib family plus individual selection, using information from more than one class of relatives, i.e. the individual itself, and its sib relatives as the other class. One could include information from parents and grandparents and half- and full-sib relatives. This work was done when Rowland was on sabbatical in Raleigh in 1980/1981. It has no application to AS.

**R. 1599. R.D. Burdon and G. Namkoong 1983: Short note: Multiple populations and sublines. Silvae Genetica 32(5–6):221–222.**
Multiple populations represent different selection criteria, to ensure that at least one such population will correspond to any future selection goal. Sublines represent replicate breeding populations that can be intercrossed, e.g. in seed orchards, to ensure completely outbred AS offspring in the future (Burdon and Namkoong 1983). Two problems:

1. Uncertainty about future economic weights of prospective selection criteria
2. The question of future inbreeding and narrowing of genetic base from selecting in closed populations

***Comment*** In the later context of the radiata BP, multiple populations would correspond to different Elites with specific end products in mind. Sublines could be reduced to a minimum of two, through CP orchards.

This is a very important paper as it formally introduces sublining and multiple populations (from which came Elites or 'breeds'). However, developments of breeding plans in 2018 probably mean these concepts will become unnecessary.

**R. 1640. R.D. Burdon 1982: Breeding for productivity- jackpot or will-o-the-wisp? Proc. North American Forest Biology Workshop, University of Kentucky, Lexington, Ke., July 1982.**
Not summarisable. A long-winded look at all angles of this question in typical Burdon exhaustive style. 'Much the biggest gains in productivity will come from appropriate combinations of species (or provenances) and management practices' (Burdon). Gains in growth rate within populations will likely be modest, while gains in quality traits, e.g. straightness, will be important.

**R. 1736. G.B. Sweet, and R.D. Burdon 1983: The radiata pine monoculture: an examination of the ideologies. NZ J. For. 28(3): 325–26.**
This is a very short paper of three paragraphs in a very defensive tone about the 'accusation' that radiata pine plantations in NZ are 'a monoculture'. Monoculture means cultivation of one crop and as an 'ideology' is defined as 'a manner of thinking, characteristic of a class or individual'.

The cultivation of RP has been assailed from an ecological standpoint as it entails major and avoidable risks, including disease, insects, climatic injury and soil degradation. The use of an exotic single species in even-aged stands and growing of repeated crops all add up to 'risks'.

Extensive reviews completed in the 1980s conclude that large-scale use of monocultures does not necessarily bring greater risks than those arising in natural forests or in mixed and/or uneven-aged stands. Evidence of risk has often been applied to monoclonal stands, but radiata stands are genetically variable. Ecological principles invoked against RP monocultures were often based on poorly defined principles.

Our defensive strategy is based on continued concentration on RP plantations of a vigorous species, well-performing in NZ, stringent quarantine, good silviculture and genetic risk spreading within the species. Research and monitoring of pathology and entomology and strength of resources of forest industries are all part of the defensive strategy. The fallback position is to have a range of 'alternative species' available with tested seed sources.

***Comment*** This an interesting 'official' reaction to the doubts that had long been expressed about heavy reliance on RP. I guess we believed the Sweet and Burdon line at the time. However, through 1950 to 1980, GTI were certainly pushing to develop other species, i.e. Douglas-fir, *P. muricata*, Mexican pines and, for different product goals, eucalypts for sawtimber (Harry Bunn et al.) and for pulp (Wilcox). There were NZ Forest Service policies for planting a substantial proportion of the estate in alternative species ('special purpose species') from the 1920s onwards, but these were not adhered to by industry like NZFP, though they did plant a resource of eucalypts from the 1970s onwards.

**D.V. Shaw and J.W. Hood 1985: Maximising gain per effort by using clonal replication in genetic tests. TAG 71:392–399.**

The most commonly cited benefit of clonal propagation is utilising non-additive genetic variance in a production population. However, the use of clonal replicates can increase the additive genetic gain relative to the non-clonal case. In the absence of much non-additive variance, use of multiple ramets does not preclude use of combined family and ramet to select good genotypes.

The relative advantage in these simulations of using multiple (versus single) ramets was rarely optimal for multistage selection. A substantial portion of the gain was from family selection. Combined index selection yields larger gains and was best for the single ramet case (i.e. non-clonal).

In many tree breeding programmes when family selection intensity must be sacrificed to avoid relatedness, substituting ramets for individual genotypes can partly compensate for the loss of intensity. As programmes mature more emphasis must be placed on within-family selection, and the use of clonal propagules in genetic tests becomes more valuable.

**R.D. Burdon 1986: Breeding long-lived perennials-frustration, temptations, opportunities. Plant Breeding Symposium DSIR 1986 Agron. Soc. NZ special pub. no. 5.**

A comprehensive look at all aspects of tree breeding from a plant breeder's perspective. Trees show long generations, long evaluation time and high cost of crop (land, investment generally). High cost of breeding, slow and irreversible maturation and the need for good breeding strategies for continued gain all complicate the picture. Good vegetative propagation is needed with shortened generation time and better screening methods. Progeny tests are very big and cumbersome. Predicting the future is difficult. Ideotype (multi- vs uninodal) is often important and selection methodology is difficult. Choice of selection criteria relative to final crop value is difficult, and there is a need for early production of improved seed in quantity. Overall, strategy is vitally important. Doing the wrong thing with the wrong species or provenance is disastrous and would need a fresh start (Burdon 1986).

***Comment*** Full of good experience and interesting angles relative to annual crops and horticulture.

**R.D. Burdon 1989: Early selection in tree breeding: principles for applying index selection and inferring input parameters. Can. J. For. Res. 19: 499–504.**
This paper elucidated the algebra of indirect selection. There are two main issues: (1) What is the optimal age for selection (advanced generation, forwards and/or backwards for seed orchards) and (2) How best to use early selection data and get highest gain per unit time?

The shorter the generation, the higher gain/unit time but the poorer the relationship of early to later performance. However, the shorter the generation time, the quicker the breeding population may 'burn up' its genetic variation, as effective population size will be run down each generation, especially if family selection is involved (Burdon 1989).

Early selection is considered as a type of indirect selection that can embrace multiple traits, each selection age data can be regarded as a separate trait in a multi-trait index.

**C.J.A. Shelbourne 1992: Genetic gains from different kinds of breeding populations and seed or plant production populations. S. Afr. For. J. 160:49–65.**
This was originally a conference paper for the IUFRO Symposium 'Intensive Forestry: The Role of Eucalypts' (Durban Sept 1991) and was about deterministic gain prediction, using the same basic equations of Namkoong et al. (1966), plus more. The scenarios were of different OP, CP and cloned breeding populations and clonal and seedling orchards and clonal forestry, under contrasting heritabilities of 0.1, 0.2 and 0.4. I also looked at cycle length and costs for different options.

Five breeding population options were covered, from solely phenotypic selection, OP families from the wild, ditto from a clonal archive, BP of full-sib families and a cloned breeding population. Twelve different production population options were covered. The paper systematically looked at the possible options, suited to eucalypts and other species. However, if the BP families were cloned, with ten clones per family with ten ramets per clone, the gain (in the $h^2$ 0.2 selection) rose from 8% to 14.1%.

It was clear from this (and other analyses) that even low intensity among-family selection in the BP, necessary to preserve status number (effective population size), was not going to give big gains each generation. Solely within-family selection combined with cloning could facilitate 'high heritability' selection done within family, to boost BP gain. Clonal selection in a clonal forestry scenario from OP families in the original OP test increased the gain dramatically, gains depending a lot on how many ramets per clone were used but would require ongoing clonal testing and then repropagation of best clones, taking another whole test cycle.

***Comment*** These methods could be applied in the NZ context, both for certain eucalypts and Douglas-fir. The cloned BP concept appears to have been adopted for radiata by the RPBC, and it is also being used in the NC State Cooperative for *P. taeda* and promoted by Fikret Isik (now associate director of their NC State-Industry Coop.)

**R. 2453. M.J. Carson, T.G. Vincent and A. Firth 1992: Control-pollinated and meadow seed orchards of radiata pine. Mass Production Technology for Genetically Improved Fast growing Forest Tree Species. Bordeaux 14–18 Sept. 1992. AFOCEL Paris 1992.**
This is a full account of the development of clonal orchards and CP orchards, including genetic gains. Special purpose breeds and seed certification integrated with breeding and plant and seed production are also covered. The role of micro- and vegetative multiplication in relation to CF is explained.

**R. 2503 R.D. Burdon 1992: Testing and selection: strategies and tactics for the future. pp. 249–60 Proc. IUFRO Conf. Oct. 9–18 1992, Cartagena and Cali, Colombia. Sponsored by CAMCORE.**
***Comment***: This could be said to be another textbook on tree breeding but which includes Rowland's latest insights. It fails to properly tackle the use of within-family selection for AS.

It includes coverage of:

- Backwards vs forwards selection (and includes a reference to Shelbourne (1992) on the use of cloning for improving gains from forwards selection within families).
- Specialisation or not, the interaction of GxE and testing strategies and different implications for forwards and backwards selection. Role of sites different from that of genotypes in GxE.
- Indirect selection systems involve different ages of assessment data, breeding for wood properties and use of genetic markers.
- Mating designs are important and need to be carefully chosen.
- Field designs and data analysis, including neighbour analysis, are the same.
- QG selection methods (various types of culling levels, index, multi-trait and combined information from relatives) are critically important (Burdon 1992).

**T.L. White, G.R. Hodge and G.L. Powell 1993: An advanced genetic improvement plan for slash pine in the southeastern United States. Silvae Genetica 42 (6): 359–370.**
The Cooperative Research Programme controls the breeding population of 933 members and effective population size of 625. The population is divided into four strata based on genetic quality, with the top stratum as an Elite population. Clones in higher strata will receive more emphasis in breeding and progeny testing. Superimposed on the strata is division into two superlines, each superline supporting 12 breeding groups (BGs)(total 24). These will allow long-term breeding while maintaining ability to create production populations of outcrossed progeny through OP or CP orchards (White et al. 1993).

Polycross tests are used to assess BVs, and full-sib families are planted in unreplicated blocks for within-family and advanced generation selection. The plan results in reducing workload from 60 to 2.5 tests or 98 ha down to 9 ha, divided horizontally into 24 groups, 39 selections each. Selections within a group are unrelated to any other BG. Two superlines each with 12 BGs will be Elites. Each BG is stratified into 3 strata (13 selections each). The more FS crosses will be GCA tested in the top breeding groups, the better.

**R. 2992 R.D. Burdon 1995: New directions in tree breeding: some questions of what we should seek, and how to manage the genetic resource. CTIA/WFGA meeting, Victoria, BC August 1995.**

Gain in productivity in crop breeding contrasts with tree breeding. Harvest index and grain yield are quite different from tree volume per ha. Tree breeders need to look at exploiting divergences between crop productivity and competitive ability. How do we translate gain from single-tree data to final whole crop production per ha.? Maybe by use of economic weights with density and volume. Gains may be different in pioneer vs climax species.

With geographic transfers in native species and as exotics, local native material may be suboptimal because of lags in adaptation and limited genetic bases. The movement of Douglas-fir north-south (not sensitive) contrasts with movement of Sitka spruce in relation to weevil.

Wood properties have high $h^2$ but are more difficult and expensive to measure.

Genetic material must be conserved as gene resources in a large BP. Tolerance of levels of inbreeding is generally unknown, including natural elimination of inbreds. This is important in sublining and subline size and may affect small breeding groups, as in new NC programme.

There is a long exposition on how to handle gene resource populations, from NZ radiata pine experience. There are problems of contamination from neighbouring plantation stands, and, as usual, there is an exhaustive consideration of anything that could happen!

The role of biotechnology in gene resource conservation has been minimal so far. Severe strains can arise from rapid advances in biotechnology in an Industry Cooperative breeding programme, and different levels and commitment to biotechnology can lead to detriment in scientific communication (Burdon 1995).

This was a long discourse, not easily summarised, and that tends to bury good points. The position in breeding of native species, especially in the PNW, is much trickier than with exotics. The important take-home lesson is that use of good provenances is very critical, especially for growth gain.

**R. 2659. L.D. Gea, D. Lindgren, C.J.A. Shelbourne and T. Mullin 1997: Complementing inbreeding coefficient information with status number: implications for structuring breeding populations. NZJFS 27(3): 255–271.**

This stochastic simulation study provided some information about the choice of size of sub-groups of breeding populations for the BP. Although the use of small disconnected groups over many generations was better at preserving status number than large groups, inbreeding in the small groups got too severe in the smallest ones, and gains were bigger in the large groups. Small groups (like the proposed NC State strategy) would not be a good long-term strategy. Small groups were fine for a strategy of very few generations (Gea et al. 1997).

Much extra gain was realised from combined index selection with more crosses per parent. Constraints on number per family selected severely reduced gains, especially at low $h^2$.

**C.J.A. Shelbourne, L.A. Apiolaza, K.J.S. Jayawickrama and C.T. Sorensson, C.T. 1997: Developing breeding objectives for radiata pine in New Zealand. Proc. IUFRO '97 Genetics of Radiata Pine, Rotorua, NZ, Dec. 1–4 1997.**
Selection indices for multi-trait and combined family and individual selection have been used in NZ tree breeding since the 1970s. Economic weights allocated in the selection indices have no real basis in the profitability or values of the traits but are important as a basis for selection to cover the wide range of end products that radiata produces (Shelbourne et al. 1997).

However, a key concept, the 'breeding objective', is needed. Following Hazel's (1943) 'aggregate genotype' = the sum of economic value x genetic value of each trait, the breeding objective should include all the main components of profitability of that production system (and is not the same as a selection index).

Breeding objective traits can be grouped as:

Forest growing, Appearance grade lumber, Structural grade lumber, Poles, Veneers, Kraft pulp and paper, and Mechanical pulp and paper.

A key separation in Solid Wood products is maximum recovery of clearwood (long internodes in unpruned logs).

Population means, between-clone, within-clone variance, broad sense heritability, heritability of clone means, range of clone means and coefficient of variation for 90 different traits, mainly product traits, are given in Shelbourne (1997) (for this conference).

**O. Rosvall, D. Lindgren and T.J. Mullin 1998: Sustainability, robustness and efficiency of a multi-generation breeding strategy based on within-family clonal selection. Silvae Genetica 47 (5/6): 307–320.**
Sustainability and robustness of genetic gain and diversity from within-family selection were stochastically analysed for the BP and PP (production population), managed according to a Swedish breeding programme for Norway spruce. This strategy is based on double-pair mating, balanced within-family selection and clonal testing within 24 populations, reproductively isolated, using stochastic simulation.

After 10 generations a linear increase in additive variance reached 11.0 sigma A, and the status number decreased from 48 to 8.1 in a BP of 48 members. Provided populations had at least 24 members, increases over 10 years in group coancestry, inbreeding and inbreeding depression were not substantial, and decreases in additive gain/generation were negligible.

Reduced additive variance and increased inbreeding had no effect on gain/generation. Clonal testing continued to be efficient throughout 10 generations down to a population size of 24 (2 sublines of 48). Clonal testing was highly effective and robust, even at low numbers of ramets and weak heritability. Low stochastic variation among replicate runs indicated high precision.

***Comment*** This is an important paper in devising a breeding strategy for radiata pine, in that the clonal testing in within-family selection is well evaluated over several generations. Most examples of cloning have been only proposals and/or involved combined clonal and family selection.

I believe using clonal testing to maximise gain from within-family selection is the way forward.

**R. 2745. R.D. Burdon 2001: Genetic aspects of risk: species diversification, genetic. management and genetic engineering. NZ J. Forestry, Feb. 2001:20–25.**

See Burdon and Miller (1995), Burdon (1982a, 2001a), Sweet and Burdon (1983).

There are five aspects of risk management:

1. Risk profiles
2. Types of risk management measures
3. Types of risk, by species
4. Genetic levels of risk for management to address
5. Specific risk management measures

Basically, this paper is about managing risk in a radiata monoculture, principally the arrival of a new disease or pest. The alternative species are themselves an element of the defences of a monoculture of radiata. Risk is a function of the probability and the seriousness of 'loss', i.e. of the health, and sustainability of the forest of radiata. Active measures of combating risk are breeding for disease or pest resistance. Risk spread also involves diversifying species and also products and markets within radiata and with other species/products, e.g. short-fibred pulp, sawtimber and veneer. Preparation of response involves introduction, testing and organisation of seed supplies of other species (note: good supplies of the Sonoma green provenance of *P. muricata* need arranging).

Risk involves biological and market aspects. Biological risk includes pests and diseases and climate, including snow, wind, temperature, drought and fire. Market risks include those of changing market preferences and needs. Past history of species diversification was largely a failure in risk avoidance and spread (*Dothistroma* on other species). Opportunity costs could be huge of using less productive and equally disease-susceptible species. Note: initial diversification into Douglas-fir, *P. nigra*, *P. ponderosa*, *P. contorta*, *P. muricata*, *P. strobus*, *P. taeda*, *P. elliottii*, *P. pinaster* and *P. patula* largely failed. Douglas-fir is the sole survivor! Eucalypts are another story.

New diseases are pitch canker and western gall rust. Hybrids with other species promise little (*P. attenuata* hybridises readily with radiata but is highly *Dothistroma* susceptible. If *P. muricata* could be made to cross with radiata this hybrid would be very useful). Radiata provenances have shown little promise except maybe Guadalupe for gall rust and Cambria for *Phytophthora*. Breed differentiation may help with market risk and *Dothistroma*. A very wide genetic base to the gene resource population may provide a source of a few 'resistant' genotypes. Risks are generally reduced by BPs of many parents, forest stands from many parents and clonal forestry stands of many clones. New biotechnology, as in genetic engineering, might help disease resistance yet carries its own limitations/risks (Burdon 2001a, b).

***Comment*** Considerations around risk are really important as a rationale for doing alternative species research.

**R. 2842. R.D. Burdon and S. Kumar 2004: Forwards versus backwards selection: trade-offs between expected genetic gain and risk avoidance. NZJFS 34(1): 3–21.**

This is an important report as it is the first indication that forwards selection can give better gains in seed orchards than backwards selection. Stochastic simulation was done of gains from a breeding population of 300 OP half-sib families, their parents selected from a population of 30,000. Backwards and forwards selections were simulated in four options:

1. 'Backwards' of 15 parents of the 300 HS families
2. 'Forwards' selection of 15 parents in 300 OP half-sib families
3. 'Forwards' selection of 15 parents within polycrosses among the same 300 parents
4. 'Forwards' selection of 15 parents from pair-cross families among the same 300 parents

*Gains were:*

$h^2 = 0.1$ and 15 parents selected for all options:

Plus trees (phenotypic) 0.59; backwards 1.90; forwards OPs 1.54; polymix 1.79; pair crosses 2.27

$h^2 = 0.2$

Plus trees (phenotypic) 0.84; backwards 2.32; forwards OPs 1.98; polymix 2.34; pair crosses 2.72

$h^2 = 0.4$

Plus trees (phenotypic) 1.18; backwards 2.81; forwards OPs 2.65; polymix 3.06; pair crosses 3.31

$h^2 = 0.6$

Plus trees (phenotypic) 1.46; backwards 3.02; forwards OPs 3.06; polymix 3.60; pair crosses 3.79

*This shows:*

Forwards gains exceed backwards for the OPs at $h^2 > 0.6$.

Forwards gains from polycrosses exceed backwards at $h^2 > 0.1$
Forwards gains from pair crosses exceed backwards at all $h^2$

This is a complicated paper and a very important one for breeding strategy. The take-home lessons for radiata breeding are very important in devising a breeding strategy for any species. For most naturally wind-pollinated species, with the fairly long periods needed for establishing clonal archives and getting CP seed (best was pair crossing), there will be reduced gains from using OP progenies for forwards selection of the new BP, but these are not massive. For $h^2 = 0.2$, gain was 1.98 for OP forwards vs 2.32 for backwards and 2.72 for pair crosses. Use of OPs for forwards selection may be compensated by much improved gain per unit time.

For seed orchards, gain from backwards selection is always inferior to gain from forwards selection via pair crosses, but breeding value estimation from CP pair crosses will take much longer than from the OP test. Much depends in this whole situation in getting CP pair crosses planted quickly, from which to backwards select a seed orchard and advance the BP. In the end, a big OP test of phenotypic selections will be excellent for roguing or reselecting orchard parents and will be a quicker yet slightly lower gain option for advancing the BP.

The bottom line for seed orchard management is that using a largish number of forwards selections will give greater gains than from backwards-selected clones. For the breeding strategist, there is not much wrong in using an OP breeding population from a gain per year perspective. Trickier is the whole question of relatedness among parents (Burdon and Kumar 2004).

***Comment*** An excellent piece of research which is also relevant for application in eucalypt and Douglas-fir breeding. It should be borne in mind that OP families from native populations may be unreliable as estimators of breeding value.

**R. 2847. R.D. Burdon, R.P. Kibblewhite, J.F. Walker, R. Evans and D.J. Cown 2004: Juvenile versus mature wood: a new concept, orthogonal to corewood versus outerwood, with special reference to *Pinus radiata* and *P. taeda*. Forest Science 50(4): 399–415.**

This includes a massive review of all the literature on variation in wood properties within the tree of a pine, particularly *P. radiata* and also *P. taeda*. An important aspect from the alternative species aspect is that it develops a new conceptual framework. Underlying this is the maturation of apical meristems in governing wood properties, even well out from the pith. Two separate concepts operate: juvenility versus maturity (in vertical/axial direction) and corewood versus outerwood (radial direction) (Burdon et al. 2004).

There is a progression of maturity up to the stem and from pith to cambium. The lower butt log contains 'juvenile corewood intergrading into juvenile outerwood'. The juvenile wood intergrades into mature wood. Further up, the stem contains mature corewood intergrading into mature outerwood. Height up the stem (echoing height growth?) determines vertical zonation.

A typical stem of radiata thus has juvenile corewood, and above this transitional corewood *inside*, and juvenile and transitional outerwood on the *outside*, say up to 5 m. Above is mature corewood with mature outerwood in decreasing quantities as one goes higher. Ten rings from the pith are the approximate extent of the juvenile core.

(Note that MFA, a good descriptor of juvenility, is highest in the juvenile corewood of the butt log, giving the lowest stiffness. Above 5 m, MFA is less, and density tends to control stiffness more than in the lower butt log.)

Extending this concept to other species and different genera works fine, but don't expect density to do the same as in *P. radiata* and *P. taeda*. Eucalypts show a different picture, yet MFA seems to match juvenility more closely than density. It would be interesting to look at Redwood (*Sequoia sempervirens*) and Chinese fir (*Cunninghamia lanceolata*) in comparison with eucalypts, *E. fastigata* and *E. nitens* for which data are available and also kauri (*Agathis australis*).

## References

LIBBY, W. J. 1969: Some possibilities of the clone in Forest Genetics Research. Genetics Lectures, Vol. I. Ed. Ralph Bogart, Genetics Institute. Oregon State University, Corvallis, Oregon.

C.J.A. Shelbourne 1969: Tree breeding methods. Technical Paper 55, Forest Research Institute, New Zealand Forest Service, Wellington.

41. C.J.A. Shelbourne and F.R.M. Cockrem 1969: Progeny and clonal test designs for New Zealand's tree breeding programme.

R. 772. C.J.A. Shelbourne 1973: Problems and prospects in the improvement of forest tree species. Proc. 2[Nd] General Congress, SABRAO, New Delhi 1973. reprinted from Indian J. Genet., 34A 1974.

R. 531. R.D. Burdon and C.J.A. Shelbourne 1971: Breeding populations for recurrent selection: conflicts and possible solutions. NZJFS 1(2):1174–193.

R. 548 C.J.A. Shelbourne 1971: Planning breeding programs for tropical conifers grown as exotics. IUFRO (section 22) Gainesville, Fla. 1971. Symposium on "Selection and breeding to improve some tropical conifers"

R. 683 C.J.A. Shelbourne 1972: Genotype-environment interaction: its study and its implications in forest tree improvement. IUFRO Genetics-Sabrao Joint Symposia.Tokyo, October 1972.

R. 1021. R.D. Burdon, C.J.A. Shelbourne and M.D. Wilcox 1977: Advanced selection strategies. Third World Consultation on Forest Tree Breeding, Canberra and Rotorua.

R. 1089. R.D. Burdon 1977: Genetic correlation as a concept for studying genotype-environment interaction in forest tree breeding. Silvae Genetica 26(5–6):145–228.

R. 1296. R.D. Burdon 1981: Generalisation of multi-trait selection indices using information from several sites. NZJFS 9(2): 145–152.

R. 1518. R.D. Burdon 1982a: Selection indices using information from multiple sources for the single trait case. Silvae Genetica 31(2–3): 81–85.

R. 1599. R.D. Burdon and G. Namkoong 1983: Short note: Multiple populations and sublines. Silvae Genetica 32(5–6):221–222.

R. 1640. R.D. Burdon 1982b: Breeding for productivity- jackpot or will-o-the-wisp? Proc. North American Forest Biology Workshop, University of Kentucky, Lexington, Ke., July 1982.

R. 1736. G.B. Sweet, and R.D. Burdon 1983: The radiata pine monoculture: an examination of the ideologies. NZ J. For. 28(3): 325–26.

D.V. Shaw and J.W. Hood 1985: Maximising gain per effort by using clonal replication in genetic tests. TAG 71:392–399.

R.D. Burdon 1986: Breeding long-lived perennials-frustration, temptations, opportunities. Plant Breeding Symposium DSIR 1986 Agron. Soc. NZ special pub. no. 5.

R.D. Burdon 1989: Early selection in tree breeding: principles for applying index selection and inferring input parameters. Can. J. For. Res. 19: 499–504.

C.J.A. Shelbourne 1992: Genetic gains from different kinds of breeding populations and seed or plant production populations. S. Afr. For. J. 160:49–6

R. 2453. M.J. Carson, T.G. Vincent and A. Firth 1992: Control-pollinated and meadow seed orchards of radiata pine. Mass Production Technology for Genetically Improved Fast-growing Forest Tree Species. Bordeaux 14–18 Sept. 1992. AFOCEL Paris 1992

R. 2503 R.D. Burdon 1992: Testing and selection: strategies and tactics for the future. pp. 249–60 Proc. IUFRO Conf. Oct. 9 - 18 1992, Cartagena and Cali, Colombia. Sponsored by CAMCORE.

T.L. White, G.R. Hodge and G.L. Powell 1993: An advanced genetic improvement plan for slash pine in the southeastern United States. Silvae Genetica 42 (6): 359–370.

R. 2992 R.D. Burdon 1995: New directions in tree breeding: some questions of what we should seek, and how to manage the genetic resource. CTIA/WFGA meeting, Victoria, BC August 1995.

R. 2621. R.D. Burdon and J.T. Miller 1995a: Alternative species revisited: categorisation and issues for strategy and research. NZ Forestry Aug. 1995

R. 2659. L.D. Gea, D. Lindgren, C.J.A. Shelbourne and T. Mullin 1997: Complementing inbreeding coefficient information with status number: implications for structuring breeding populations. NZJFS 27(3): 255–271.

C.J.A. Shelbourne, L.A. Apiolaza, K.J.S. Jayawickrama and C.T. Sorensson 1997: Developing breeding objectives for radiata pine in New Zealand. Proc. IUFRO '97 Genetics of Radiata Pine, Rotorua, NZ, Dec. 1–4 1997

O. Rosvall, D. Lindgren and T.J. Mullin 1998: Sustainability, robustness and efficiency of a multigeneration breeding strategy based on within-family clonal selection. Silvae Genetica 47 (5/6): 307–320.

R. 2745. R.D. Burdon 2001a: Genetic aspects of risk: species diversification, genetic. management and genetic engineering. NZ J. Forestry, Feb. 2001:20–25.

R. 2842. R.D. Burdon and S. Kumar 2004: Forwards versus backwards selection: trade-offs between expected genetic gain and risk avoidance. NZJFS 34(1): 3–21.

R. 2847. R.D. Burdon, R.P. Kibblewhite, J.F. Walker, R. Evans and D.J. Cown 2004: Juvenile versus mature wood: a new concept, orthogonal to corewood versus outerwood, with special reference to *Pinus radiata* and *P. taeda*. Forest Science 50(4): 399–415

R. 2745. R.D. Burdon 2001b: Genetic aspects of risk: species diversification, genetic. management and genetic engineering. NZ J. Forestry, Feb. 2001:20–25.

W.J. Libby 1964: Clonal selection and an alternative seed orchard scheme. Silvae Genetica13; 32–40.

G . Namkoong, E.B. Snyder and R.W. Stonecypher 1966 Heritabilites and gain concepts for evaluating breeding systems such as seedling orchards. Silvae Genetica 15(3):76–84.

R.W. Allard. 1960 Principles of Plant Breeding. John Willey and Sons Inc., New York.

Burdon, R. D. and Low CB. 1973: Effects of site on expression of cone characters in radiata pine. New Zealand Journal of Forestry Science 3 (1): 110–19.

# Chapter 13
# Radiata Breeding

King et al. (1988) nicely reviewed the radiata breeding programme after the Development Plan at the Wagga Wagga, NSW meeting. Sue Carson (1989–1991) reported results of the disconnected diallel trials at ten sites in an in-depth study of genotype × environment interaction. Her conclusions on optimal number of progeny test sites were useful and applicable in the breeding context. This is a good general guide to coping with GxE in the other AS breeding programmes.

**I.J. Thulin 1957: Application of tree breeding to New Zealand forestry. NZ Forest Service Tech. Pap. 22 (1957).**
This is a technical paper, normally published by NZFS head office. It was also published in the proceedings of the (British) Comm. For. Conf. Aus. and NZ 1957.

This is the first paper of Ib's that sets out his 'philosophy' of tree breeding and genetic improvement generally. He rejects the idea of provenance variation in *P. radiata* because of its limited distribution in California and regards the trees selected in NZ plantations as the main material for selection. 23,000 acres were searched in the Kaingaroa plantations to select 15 trees. Vegetative propagation via cuttings was regarded as satisfactory up to ortet age 12 years, but clones would have an adult habit. The clonal archive was the key element for breeding and for vegetative propagation.

The first two 10 ha. clonal seed orchards of radiata pine were under establishment. Clonal and progeny tests were seen as necessary for genetic testing of orchard clones.

Provenance research was essential for other species which had wide geographic ranges, and provenance testing was a necessary first step. There were reports by Egon Larsen on seed collections in Oregon and California of similar latitude and climate to NZ. Ib discussed provenance problems in Douglas-fir, *P. ponderosa*, *P. contorta*, *P. nigra* and *P. pinaster*. He noted 50,000 acres of Douglas-fir had been successfully planted in the NI and SI and reported the establishment of the 1957 series of Douglas-fir provenance trials, mainly of commercial seedlots from northern regions of the West Coast of the USA. Plans were also outlined for *P. ponderosa*,



C. J. A. Shelbourne, M. Carson, *Tree Breeding and Genetics in New Zealand*,
https://doi.org/10.1007/978-3-030-18460-5_13

*P. nigra*, *P. pinaster* and *P. contorta*. By 1957, Ib had a good mental picture of the geographic trends in provenance variation in all these species, based on observations of NZ stands and on early results in trials (1927 Moore study for *P. ponderosa*, published 1944, was the first).

He proposed new species/provenance trials with a few provenances for several species. He ended with measures for improvement from seed orchards and seed stands and provenance trials to ensure seed is imported from good provenances.

**R. 374. I.J. Thulin 1969: Breeding *Pinus radiata* through seed improvement and clonal afforestation. Second World Consultation on Forest Tree Breeding, Washington, 7–12 Aug. 1969.**

(Note: there is a 12-year gap between Ib's 1957 first paper and this one).

The introduction focuses on poor stem quality of radiata, strong basket whorls, steep branch angles and crooked internodes all requiring the planting of large numbers to get sufficient good-formed trees for a final crop.

He dates the start of tree breeding to 1953 with intensive plus-tree selection and clonal seed orchards and forecasts 10,000 lb of seed by 1976. An assessment of two 11-year-old CP tests in 1967 (one of my first jobs) revealed improvement of 63% in stem straightness, 45% in branching characteristics and 14% in volume over a bulk seedlot.

He then describes cutting propagation, with 80% of all clones rooted up to 12 years old. 'Plantable rooted cuttings from trees up to 15–17 years old can be produced in commercial quantities at a cost of $20/1000 vs $7/1000 for seedlings'. Ib really stuck his neck out on this when the oldest cutting-propagated trials were only aged 2 years from planting. He saw that maturation was giving a bonus of good form and few defects. He went on from there to 'Development of a clonal breeding programme' with selection of clones on total genetic variance and repropagation of best clones.

The '266' clones were selected and propagated in 1966 and planted in 1968. Growth of 1 + 1 cuttings had been good, based on 5- to 6-year-old ortets, so this probably added to Ib's overconfidence. I believe we quickly realised that an extensive clonal forestry programme, as described, wasn't going to happen, but we continued to try with such series as the '870', '871' and '873' propagated from CP seedlings of 4, 5 and 7 years from seed.

He then describes the '268' programme, with a bias towards selecting large numbers of plus trees, propagating all by cuttings and producing large numbers of cuttings for bulk repropagation, and this is from ortets 12–17 years old! This was written in December 1968 when the '268' cuttings had only been set in nursery beds a few months previously (and the seed of ca. 600 progenies sown in September), and he concluded that we got 80% strike. He then went on to tabulate how 500 clones could be bulked up over years from repropagation, with parallel clonal tests reducing the number of clones being propagated and planted commercially. In retrospect, it was all dreadful 'pie in the sky', and I don't remember sighting the paper before the meeting. He used my predicted gains from tree breeding methods (a highly abbreviated version was presented at this consultation) to reason that clonal forestry via clonal testing and afforestation in one cycle of selection would provide greater improvement than seed orchards, with the advantages of a 'quick, effective and relatively cheap method of improvement, good for low $h^2$ traits, disease resistance, wood quality, and adaptability to cold and poor nutrition sites'. However, he did add that one still needed some form of recurrent selection and/or interspecific hybridisation to give improved populations for further clonal selection.

***Comment*** As I remember things, the cutting propagation of the '268' ortets was sort of experimental (their age was against it), and by winter 1969, it was clear that only a small fraction of the clones would produce enough cuttings to plant a small clonal trial, which was done, only at Cpt. 1350. Their grafting was done so that there would be scion material of all clones available for seed orchard grafting and for archives for control pollination, also at Cpt. 1350.

This all indicates a very 'gung-ho' attitude towards clonal forestry, by Ib and to a lesser extent the rest of us. However, the whole '268' progeny testing programme and clonal orchards were realistic and soundly based, and this was soon backed up by Rowland Burdon, and later by Mike Wilcox, in applying early backward and forward index selection.

**61. C.J.A. Shelbourne, I.J. Thulin and R.M.H.C. Scott 1972: Variation, inheritance and correlation amongst growth, morphological and wood characters in radiata pine.**

The three OP trials at Gwavas, Golden Downs and Berwick were assessed in 1967 when aged 12 years from planting. The 29 OP progenies in 3 reps of 20-tree row plots at each site provided the first genetic parameter estimates for a wide variety of traits of 12-year-old material. Site x family interaction was not significant over the three sites, Gwavas, Golden Downs and Berwick. Narrow sense $h^2$ were ranked: number of branch clusters 0.86, wood density 0.56, height 0.54, height to first stem cone 0.41, number of stem cone clusters 0.38, taper at half height 0.29, branches/cluster 0.26, branch diameter 0.32, volume 0.20, DBH 0.15, branch thickness 0.11, crown radius 0.11 and branch angle 0.08.

***Comment*** The most important phenotypic and genetic correlations were between branch cluster number and growth rate and branch diameter and stem coning. Selection of multinodal trees should result in multinodal offspring with increased growth rate and reduced branch diameter, also with increased stem coning. This was a classic heritability study which gave us good estimates of genetic parameters, unfortunately delayed by 4 years.

This study of radiata pine gave me some confidence in OP tests for family ranking and for genetic parameter estimates in RP and other conifers. It was a major study that took far too long to analyse, mainly because we had no direct access to a computer at that time.

**C.J.A. Shelbourne 1970: Genetic improvement in different tree characters of radiata pine and the consequences for silviculture and utilisation. Proc. FRI Pruning and Thinning Symposium, Rotorua March 16 1970. Paper 13: 44–58.** This paper summarised results from 11-year-old CP tests at Cpt. 1038 Kaingaroa and at Golden Downs. The mating design was an aborted NCII factorial in which '850'-19 and '850'-55 were used as male testers, plus '850'-20, a fastigiate-formed genotype. '850' clones were used as females.

A bulk unselected ex-Kaingaroa seedlot was included which allowed some estimates of genetic gains. By age 11 with 7–8 foot spacing and 20-tree row plots, the inferior control lot (and others) was suffering from competition which must have exaggerated gains. An average of all crosses was superior to the control in bole straightness, butt sweep, forking, ramicorns, number of branch clusters and acceptability. The best two families were highly superior in every trait except branch diameter and branch angle which were still better than the bulk control lot.

This was the first concrete 'good news' for the breeding programme, which till then had relied on faith in intensive selection of the plus trees. The emphasis in this paper (for the Pruning and Thinning Symposium) on defining the effect of different characters on end products was a precursor to the Douglas-fir sawing and stress grading study.

**M.D. Wilcox, C.J.A. Shelbourne and A. Firth 1976: General and specific combining ability in eight selected clones of radiata pine. NZJFS 5 (1975/76).** Eight selected seed orchard clones of radiata pine were mated together in a factorial design with four clones as females and four as males and planted in a single-tree plot (STP).

At age 5 years, volume, straightness, branching and resistance to Nemacylus needle-cast were all showing high GCA effects, but height, DBH and volume also

**Fig. 13.1** Full-sib family 100 tree blocks, p. 1968. Probably 55 × 96 on right

showed some SCA. This experiment was done at Cpt. 1350 and included OP orchard progenies of these same parents, all planted in a single-tree layout, the first of its kind, which was a trail blazer for the use of STPs later.

**R. 1019 J.M. Harris and M.D. Wilcox 1976: Priorities for tree selection in radiata pine. XVI IUFRO World Congress, Oslo 1976.**
This covers the history of tree breeding for radiata to 1976: '850' selection in the '50s and later '268's. It quotes Wilcox, Firth, Low and McConchie 1975 GTI report 78 on the first assessment of the '268' OP tests and gives heritability values for combined forward selection (among and within families) for volume 0.40, stem straightness 0.38, branch quality 0.53 and wood density of 0.79. This was the first application of index selection in NZ.

It also quotes Harris, James and Collins 1976/7 in JZJFS 'Case for improving wood density' and covers a stress grading study like the Douglas-fir one. In the past RP was harvested at about age 45, but reduced age of future harvest will have reduced density and lowered strength. Lower proportions of outerwood may be offset by pruning and by tree breeding. Machine stress grading is an essential element of the future programme (Fig. 13.2).

**R.D. Burdon and C.B. Low 1977: Variation in needle characters of *Pinus radiata* from mainland California. NZJFS 7.**
Cambria population needles were longer and Ano Nuevo were shorter (Fig. 13.3).

**C.J.A. Shelbourne, R.D. Burdon, M.H. Bannister and I.J. Thulin 1979: Choosing the best provenances of radiata pine for different sites in New Zealand. NZ J. For. 24: 288–300.**
This includes the results of two simple provenance experiments, one planted in 1955 and the other in 1968. The 1955 trial was planted at Gwavas and Golden

**Fig. 13.2** Single-tree plots of 850 crosses, showing a range of performance within one rep. Long mile demonstration area

**Fig. 13.3** The best tree in the block, 850-55 is one parent. Single-tree plots in long mile demonstration area

Downs and included two seedlots from Monterey (Huckleberry Hill and Pt. Lobos) and one lot from Cambria and OP progenies of '850'-7 and '850'-55.

There were 100-tree plots at Gwavas and 196-tree plots at Golden Downs (typical of early provenance trial design). At Golden Downs, ranking of provenance means adjusted for stocking was OP 55, Huckleberry Hill, Pt. Lobos and Cambria. At Gwavas ranking was OP55, Huckleberry Hill, Pt. Lobos and Cambria equal. (Plot number per site varied from 2 to 4.) The 1968 trials were based on surplus stock from the big provenance-progeny trials in Kaingaroa and were planted at Gwavas, Santoft and Golden Downs. They included a mix of Kaingaroa and Nelson NZ populations as well as Monterey, Ano Nuevo and Cambria provenances and the Guadalupe Island population.

There were some rank changes with site. At Gwavas the NZ populations were ahead in DBH and volume and not much different from the three mainland Californian populations. Guadalupe was much slower grown. At Santoft, Nelson-Kaingaroa were barely ahead of Monterey and Cambria, and Ano Nuevo was very slow grown. At Golden Downs, Monterey was clearly the fastest, Nelson-Kaingaroa were the second, and Cambria and Guadalupe were well down. The conclusion seems to be that in the small sample of sites tested, relative performance of the provenances and NZ land races can vary considerably.

**C.J.A. Shelbourne and C.B. Low 1980: Multitrait index selection and associated genetic gain of *Pinus radiata* progenies at five sites. NZJFS 10.**
In 1971 220 of 588 OP families of the '268' series originally planted in 1969 at Kaingaroa, Gwavas and Waimihia were planted, with an additional 80 ex SI families, at Woodhill, Golden Downs and Otago Coast, using a sets in reps design with 30 families per set. Components of variance were estimated for each site to calculate within site multi-trait selection indices. An overall site index combined a total of 17 site traits using the Burdon's method. The families selected agreed with those selected using summed rankings. The '268' families proved to be an adaptable group with superior growth to the local SI selections (Fig. 13.4).

**M.H. Bannister and M.H. Vine 1981: An early progeny trial in *Pinus radiata* 4. Wood Density. NZJFS 11 (1981).**
There were 26 OP families grown at Pigeon Valley, Nelson. Samples were divided into five ring segments; $h^2$ per segment varied from 0.51 to 0.72. Pith to bark gradients differed between families. There was no correlation of density with other traits.

**Fig. 13.4** Family variation in ten-tree row plots of the '268' OP test at Cpt. 1350 Kaingaroa

**M.D. Wilcox 1982: Genetic variation and inheritance of resistance to Dothistroma needle blight in *Pinus radiata*. NZJFS 2.**

Three separate studies showed that NZ populations of *P. radiata* possess useful quantitative variation in resistance to infection by *Dothistroma*. Heritability was high enough to allow effective selection and breeding for resistance.

**Eldridge, K.G. 1982: Genetic improvements from a radiata pine seed orchard. NZJFS 12.**

Three yield trials, 10–12 years old, have shown that seed from the first seed orchard at Tallaganda produced 20% more volume and twice as many trees of excellent stem and branch quality than the control seedlot.

**Genetic gain trials planted 1978 and 1979. No reports available (Fig. 13.5).**

**D.V. Shaw and J.W. Hood 1985: Maximising gain per effort by using clonal replication in genetic tests. TAG 71:392–399.**

The most commonly cited benefit of clonal propagation is utilising non-additive genetic variance in a production population. However, the use of clonal replicates can increase the additive genetic gain relative to the non-clonal case. In the absence of much non-additive variance, use of multiple ramets does not preclude use of combined family and ramet to select good genotypes. The relative advantage in these simulations of using multiple (versus single) ramets was rarely optimal for multistage selection. A substantial portion of the gain was from family selection. Combined index selection yields larger gains and was best for the single ramet case (i.e. non-clonal).

**Fig. 13.5** Genetic gain trials: large-plot pulpwood regime with no thinning. p. 1978

In many tree breeding programmes when family selection intensity must be sacrificed to avoid relatedness, substituting ramets for individual genotypes can partly compensate for the loss of intensity. As programmes mature, more emphasis must be placed on within-family selection, and the use of clonal propagules in genetic tests becomes more valuable.

**A.C. Matheson and D. Lindgren 1985: Gains from the clonal and clonal seed orchard options compared for tree breeding programmes. TAG 71 (2)242–249.**
Predicted gains were compared from selection of clones from a plantation (without further clonal testing and selection) with a seedlot from a clonal seed orchard of the same clones. The extra gains from the CF option were only due to the shorter time between selection and field planting of the CF clones compared with seed orchard seedlings.

**R.D. Burdon and M.H. Bannister 1985: Growth and morphology of seedlings and juvenile cuttings in six populations of *Pinus radiata* NZJFS 15.**
Comparison of seedlings and cuttings 4 years from seed, taken at year 3, showed that cuttings taken 50 cm above the root collar had maturation but outgrew seedlings. Cuttings had thinner bark, better straightness, branch angle and branch habit.

**S.D. Carson and M.J. Carson 1986: A breed of radiata pine resistant to Dothistroma needle blight.**
*Dothistroma* needle blight affects large areas of PR in NZ. It is controlled by a costly annual spray programme. Genotypic variability in resistance has been quantified. $h^2$ is moderately high, and field screening for resistant trees works. Genetic gains in resistance can be expected from progeny testing and recurrent selection. Work is under way to develop a resistant breed while maintaining current levels of other traits (Figs. 13.6, 13.7, and 13.8).

**M.J. Carson 1986: Control pollinated seed orchards of best general combiners: a new strategy for radiata pine improvement. Proc. Plant Breed. Symp. Lincoln, New Zealand.**
Results from a disconnected diallel of the '875's at Onepu and Cpt. 327 Kaingaroa indicate that GCA effects are much more important than SCA for DBH, straightness, and *Dothistroma* resistance. These support the use of manually isolated control-pollinated clonal seed orchards of best general combiners for all traits (a control-pollinated seed orchard implies a hedged, isolated with cellulose sleeves and manually pollinated orchard). Using mass CP and vegetative multiplication, maximum gains can be achieved by selecting for parental GCA. This was done through OP tests in the past and will most likely be done with a polycross or NCII in future. New CP orchards will allow higher-intensity parental selection with no pollen contamination. Special purpose breeds are easily produced on demand from CP orchards. Incidentally, Burdon found no evidence of SCA in an NC Design 2 trial with four testers.

**Fig. 13.6** Severe Dothistroma defoliation on 3-year-old PR

**Fig. 13.7** The pollination operation at Amberley

**Fig. 13.8** Ruth McConnochie and 5+-year-old grafts at Amberley

**S.D. Carson and M.J. Carson 1986: A breed of radiata pine resistant to *Dothistroma* needle.**
*Dothistroma* needle blight resistance has been quantified and $h^2$ is moderately high. Field screening for resistant trees works. Genetic gains in resistance can be expected from progeny testing and recurrent selection.

**C.J.A. Shelbourne, R.D. Burdon, S.D. Carson, A. Firth and T.G. Vincent 1986: Development plan for radiata pine breeding. Forest Research Institute, Rotorua, New Zealand.**
Section 1 Historical introduction and Section 2 Review of breeding programmes are important in detailing the whole structure of the different clone selection series, '850', '268', '875', '880', etc. Section 3 Breeding strategy is very important. There was great excitement about the impact of CP orchards on breeding strategy. As reported by Shelbourne, Burdon et al., 'the successful adoption of the new technology (control pollinated orchards) therefore promises to be a quantum leap forward in the effectiveness of applied tree breeding. The new technology…promises to allow much greater specialisation of different breeds…down to specific crosses. The limitations will lie in the scope of progeny testing of orchard clones…gains must inevitably be greatly increased over those of a general-purpose breed'. (And one should add 'over OP clonal seed orchards'.) Coping with GxE and the needs of selection criteria on different sites, e.g. sands, Auckland clays, etc. is potentially solved by CP orchards.

***Comment*** Note that backward selection of the CP orchard parents was still seen as necessary. We were not yet alert to the losses in time of using backward selection in all aspects of the breeding and orchard programmes, as opposed to forward selection.

The future success of CP orchards was correctly seen as closely tied to the development of cutting propagation from nursery stools, with resulting rooted cuttings. Time delays from different means of GCA testing, using OP, CP on original ortet and CP on hedged clones, were given as 1, 4 and 8 years. These numbers need more emphasis. OP obviously is quickest, but isolation and pollination by climbing the selected tree, which needs to be at least 8 years from planting, is an option, especially for mating designs like SPM and polycrossing. This is the first time the 'time' dimension has been mentioned.

Under 'breeding populations' the whole problem of the size of the breeding population and its composition is addressed. The main conflict is between getting maximum gain using within- and among-family selection and maintaining a breeding population with solely within-family selection. Intensive selection and restriction of the number of clones in the production population was not a problem. The best strategy is to use within-family selection for the breeding population and between-family or between-parent selection for the orchards.

***Seed Producing Population***

Current clonal orchards were selected by backward selection from progeny test results on parent GCA. The main problem for conventional orchards was getting enough rooted cuttings from a few clones, especially for conventional orchards (assuming serious graft incompatibility). CP orchards solve most of these problems, including maximising genetic gains from those clones, dependent on good GCA estimates from progeny tests. CP orchard technology allows greater specialisation of breed to site (but this may not be needed; see Sue Carson on GxE and the disconnected diallel).

One can argue now, through stochastic simulation, that forward selection in CP orchards can give more gain per year than backward selection (see Burdon and Kumar 2004).

Early estimation of GCA by climbing and pollinating ortets or using OP seed is important for CP orchard management. OP tests are a satisfactory means of estimating GCA for roguing and reselection of parents. They could be used for AS but give reduced gain per generation or in a family block situation.

***Time Delays* (Table 13.1)**

*Selection Criteria*

Economic weightings of selection criteria were a problem but SILMOD is a means of revealing more. Still, this is a major problem, deciding what economic weightings for the various traits to give an index, against a background of genetic parameters.

**Table 13.1** Time delays

| | |
|---|---|
| Clonal orchard (after selection) | Cuttings 2 years, grafts 1 year |
| Clonal orchard establishment | From hedged archives 5–7 years |
| OP test establishment | 1 year |
| CP test establishment (mating on ortets) | 4 years |
| CP test establishment (mating on hedges) | 8 years |

***Mating Designs***

OP and polycross mating of parents are both good for estimating GCA. However, the polycross suffers the same disadvantages of cost and time as full-sib designs. OP tests have been used for AS in the past (e.g. '875', '880's) and will be used for a supplementary NZ BP.

A seedling tree can only grow in one place, while its family can be planted on several sites. In a background of planting single-family blocks, this is particularly true for clones: the possibilities for using rooted cutting clones, planted on several sites, for forward selection (within families) remain to be explored.

The disconnected factorial and diallel designs are good for AS but no good for GCA estimation. Sublining will reduce inbreeding in the BP. Some immediate costs are incurred for extensive 'on-ortet' pollination, though they are a lot less than archive establishment and CP. Note that SPM is more economical and gives as good gains for AS as disconnected diallel and factorial but still requires resources for climbing ortets but only half the crosses.

Second-generation selections have been made since 1974 ('875's) on family and individual performance, but selection on family means entails reduction in status number. Clonal testing of individuals within full-sib families using juvenile cuttings is potentially the most precise means of selecting individual genotypes available, but practical application still has to be developed.

Rowland Burdon's Section 4 Gene resource management plan starts with: 'It behoves breeders...to put together a comprehensive set of genetic defences of which the gene resource is one element'. 'Genetic defences' means 'available genetic variability' and 'the means of using it for minimising crop vulnerability.' Rowland separates 'short-term defences' and 'long-term defences' for future crops.

This section sets out detailed plans, requiring quite massive inputs for gene conservation, and flow charts for the Genetic Survey populations and the Firth-Eldridge collection.

1. For gene resource conservation, from the original Genetic Survey experiment, select the 100 best phenotypes per population, collect seed and bulk OP families and plant each population bulk.
2. Cross best Guadalupe hybrids with NZ population.
3. Firth-Eldridge: country-wide provenance testing (in progress) and block plantings (done).
4. Conserving NZ land race stocks: survey existing stands of regional seedlots and track to ultimate felling.

*Sue Carson's Section 5* on 'Breeding for disease resistance' was a set of plans for the *Dothistroma*-resistant breed and for coping with *Cyclaneusma* in the main breeding programme. A good chapter but very specific to those radiata diseases. Also covered are genetic parameters, GxE and regionalisation from the disconnected diallel at 11 sites. Refer to S.D. Carson 1989; S.D. Carson 1990: S.D. Carson 1991.

*Gerry Vincent's Section 6* describes how parent clones of the '268' selection were handled. First selected in 1969, in 1974 cuttings of some untested clones of the series were planted at the Kaingaroa orchard and progressively reduced to 65 clones by 1984.

**Fig. 13.9** Gerry Vincent doing some pollination on a very small graft

The new Amberley CP orchard was a converted conventional OP orchard of 106 clones, mostly '268' and some '668' and '850's. In the first CP programme in 1986, it was rogued to the best 42 clones, and the best 15 were used in the pollen mix (Fig. 13.9).

This account highlights the bind orchard managers and breeders were in, trying to get orchard areas planted quickly of new selected clones. Shortages of scion and cutting material from archives and the necessary wait for test results from the progeny tests slowed things down and constrained the whole operation. This would be minimised by using OP tests and can take a long time with CP tests, maybe 8 years plus another 8 for tests to grow. Part of the 'time' problem was the use of the long mile for CP. Frosts, etc. delayed effective production of CP seed by many years compared with Amberley (this underlines the problem of extended generation time of using backward selection for orchard clones).

Gerry highlights the interaction between seed of the genetic quality that was available and annual commercial nursery usage. Seed can be collected from the best GCA clones in the orchard when seed is in good supply but demand is low. Use of precision sowers in nurseries reduced seed consumption/requirements.

CP seed orchards have become a commercial reality, and the area of these is likely to increase in the future.... Assuming a planting programme of 40,000 ha with...1000 trees/ha and 15,000 plants/kg of seed, 2700 kg of seed is required each year. With multiplication × 20 of cuttings per seed, only 135 kg seed is needed.

***Section 8. Research and Development (Tony Shelbourne)***

Top priority five problems were:

1. Defining selection criteria – what to select for? Higher wood density? Long and short internode breeds?
2. Development of mass vegetative propagation systems for seed from CP orchards. Effects of maturation/ageing?
3. Development of commercial CP orchards. Establishing a commercial CP hedged orchard enterprise can double gains from best general combiners in OP orchards.

4. Determine regionalisation of BP and/or orchard programme and importance of GxE.
5. Provenance adaptation across NZ. 1980 Firth-Eldridge collection and its wide planting will give answers.

Additional difficulties included the following:

6. Orchard propagation bottleneck. Delays and difficulties in propagating orchard clones older than 6 years.
7. Early screening.
8. Breeding strategy
9. How big should the BP be?
10. Weighting of among-family versus within-family selection.
11. Sublining or not.
12. Fitting existing series into a BP.
13. Gain prediction and measurement (gain trials). Still worried about non-additive variance (SCA) versus additive.
14. Relationship of STP results and final crop results. Large plots versus STPs.
15. Economic weights.
16. Selection methodology.

***Section 9. Development of BP-Requirements***

- Large effective population size
- Maintaining a limited number of selection criteria (Fig. 13.10)

**Fig. 13.10** Breeding population size, number of families and number of trees per family

***Controlling Inbreeding***

- CP orchards only need two sublines which makes it easier to integrate the numerous clone series in the BP.
- Selection for GCA in BP? No! but for PP is very necessary, achievable through OP, polycross or NC2.
- Among-family selection should be reduced or eliminated in BP, with increased within-family selection in BP.
- CP on standing ortets is feasible, increases gain/generation and is much quicker than in a hedged clonal archive.
- Designs: Within-family selection in blocks of single families. Need GCA estimates via OP or polycross. But selection within family has low $h^2$ (resolvable by cloning) (Fig. 13.11).

***Use of Cloning in Genetic Selection***
This approach will rely on within-family forward (clonal) selection for selecting future members of the BP. Forward selection *within* BP families will advance both BP and orchards. Otherwise selection of best trees in single-family blocks of seedlings will provide the new BP.

The PP will be selected 'backward' from GCAs from the polycross or from OP families. The use of cloning within families will improve the accuracy of 'forward' selection within families, selecting the clones for both BP and PP. How to integrate

**Fig. 13.11** Effects of extreme inbreeding: self versus outcross

the sometimes-haphazard breeding work of last three decades into a strategy for two more decades? Two key features are complementary mating designs and CP orchards. There is also the exciting new possibility of cloning for forward selection within families.

(There are) eighteen published references (4 Burdon, 3 Carson, 5 Shelbourne, 3 Namkoong, 3 Wilcox) and 35 unpublished or limited circulation references (5 Burdon, 5 Carson, 14 Shelbourne, 11 Wilcox).

**J.N. King, R.D. Burdon, A. Firth, G.R. Johnson, C.B. Low, C.J.A. Shelbourne and T.G. Vincent 1988: Advanced-generation breeding of radiata pine in New Zealand. Ninth Australian Plant Breeding Conference, Wagga Wagga, NSW. June 1988: 263–264.**

This was a valuable summary paper, probably from a verbal delivery, of 2 years later than the 1986 Development Plan and featured current thinking around the 'Big Bang':

1. Hierarchical structure (the pyramid)
2. Seed production in CP orchards with vegetative multiplication of CP seedlings.
3. Separate testing and field designs for crosses for advancing the BP and for GCA testing.
4. Breeding population with two crosses per parent and some assortative mating by grouping into three
5. Classes, more crosses for top-ranked parents (which unfortunately didn't eventuate).
6. CP orchards. Many advantages, particularly only two sublines needed for out-crossing in the orchard.
7. Family blocks for within-family selection.
8. Sublining, but with CP orchards, no need for more than two.
9. Multiple-population breeding (assortative mating in 'product'- or 'site'-related groups). Elites?
10. 'Systematic' approach with different subline groups spread over years.
11. GCA testing with female tester design, STPs, 3–4 sites, cross-referenced control lots to connect different subline groups.

Note: Separation of GCA testing and population advancement crossing and assortative mating.

***Comment*** This stated nicely where radiata breeding strategy had got to by 1988.

## References

I.J. Thulin 1957: Application of tree breeding to New Zealand forestry. NZ Forest Service Tech. Pap. 22 (1957)

R. 374. I.J. Thulin 1969: Breeding *Pinus radiata* through seed improvement and clonal afforestation. Second World Consultation on Forest Tree Breeding, Washington, 7-12 Aug. 1969.61.

C.J.A. Shelbourne, I.J. Thulin and R.M.H.C. Scott 1972: Variation, inheritance and correlation amongst growth, morphological and wood characters in radiata pine

C.J.A. Shelbourne 1970: Genetic improvement in different tree characters of radiata pine and the consequences for silviculture and utilisation. Proc. FRI Pruning and Thinning Symposium, Rotorua March 16 1970. Paper 13: 44–58

M.D. Wilcox, C.J.A. Shelbourne and A. Firth 1976: General and specific combining ability in eight selected clones of radiata pine. NZJFS 5 (1975/76

R. 1019 J.M. Harris and M.D. Wilcox 1976: Priorities for tree selection in radiata pine. XVI IUFRO World Congress, Oslo 1976.

R.D. Burdon and C.B. Low 1977: Variation in needle characters of *Pinus radiata* from mainland California. NZJFS 7.

C.J.A. Shelbourne, R.D. Burdon, M.H. Bannister and I.J. Thulin 1979: Choosing the best provenances of radiata pine for different sites in New Zealand. NZ J. For. 24: 288–300.

C.J.A. Shelbourne and C.B. Low 1980: Multitrait index selection and associated genetic gain of *Pinus radiata* progenies at five sites. NZJFS 10

M.H. Bannister and M.H. Vine 1981: An early progeny trial in *Pinus radiata* 4. Wood Density. NZJFS 11 (1981).

M.D. Wilcox 1982: Genetic variation and inheritance of resistance to Dothistroma needle blight in *Pinus radiata*. NZJFS 2

Eldridge, K.G. 1982: Genetic improvements from a radiata pine seed orchard. NZJFS 12

D.V. Shaw and J.W. Hood 1985: Maximising gain per effort by using clonal replication in genetic tests. TAG 71:392–399

A.C. Matheson and D. Lindgren 1985: Gains from the clonal and clonal seed orchard options compared for tree breeding programmes. TAG 71 (2)242–249

R.D. Burdon and M.H. Bannister 1985: Growth and morphology of seedlings and juvenile cuttings in six populations of *Pinus radiata* NZJFS 15.

S.D. Carson and M.J. Carson 1986a: A breed of radiata pine resistant to Dothistroma needle blight.

M.J. Carson 1986: Control pollinated seed orchards of best general combiners: a new strategy for radiata pine improvement. Proc. Plant Breed. Symp. Lincoln, New Zealand.

S.D. Carson and M.J. Carson 1986b: A breed of radiata pine resistant to Dothistroma needle cast.

C.J.A. Shelbourne, R.D. Burdon, S.D. Carson, A. Firth and T.G. Vincent 1986: Development plan for radiata pine breeding. Forest Research Institute, Rotorua, New Zealand.

J.N. King, R.D. Burdon, A. Firth, G.R. Johnson, C.B. Low, C.J.A. Shelbourne and T.G. Vincent 1988: Advanced-generation breeding of radiata pine in New Zealand. Ninth Australian Plant Breeding Conference, Wagga Wagga, NSW. June 1988: 263–264.

R. 2842. R.D. Burdon and S. Kumar 2004: Forwards versus backwards selection: trade-offs between expected genetic gain and risk avoidance. NZJFS 34(1): 3–21.

S. D. Carson, Breeding for Resistance in Forest Trees-A Quantitative Genetic Approach 1989–1991.

# Chapter 14
# Breeding Strategy

In 1968, a completely different kind of plus-tree selection of 600 trees (268s) was carried out in 12–17-year-old plantations in Kaingaroa at a much-reduced rate of 1 tree per 3 acres. Open-pollinated progenies of these 588 trees were planted at three sites. Five years later the trials were assessed with the main purpose of selecting the best of the original 588 parents for a future orchard. This was a 'backwards' selection operation using the progeny performance to 'backwards' select the best parents.

This was the start of an existential fight between backwards and forwards selection. About 100 trees were selected from the better families in what was 'forwards' selection and control-mated (on standing trees). However, the resulting CP families were used only in backwards selection to identify the best 875 parents, that is, those second-generation trees on which the mating was done. They were suited to form the third generation but were never used to advance the breeding population. The concept of a breeding population was now existant, but breeders (including me) did not think in terms of forwards selection to nominate the future breeding population. Thus, the breeding population was forgotten in the search for the best seed orchard parents.

And, so, the conflict between backwards and forwards selection went on. Another concept that breeders were generally unaware of was the genetic variance in the population for any trait. Shaw and Hood in 1985 pointed out the value of within-family selection which did not reduce genetic variance in the population, and clonal replication of seedlings within those families made selection much more precise.

In the Development Plan of 1986, there was a realisation that within-family selection was valuable (no reduction in genetic diversity in BP) and that cloning of the trees within the family would give high gain. However, the rooting of young seedlings was not yet operational. The control pollination of the mostly backwards-selected parents and the planting of the Big Bang unfortunately met the caterpillar disaster of Helicoverpa.

In spite of the Development Plan, there had been a strong tendency to use backwards selection, not just for creating seed orchards, where intense backwards



C. J. A. Shelbourne, M. Carson, *Tree Breeding and Genetics in New Zealand*,
https://doi.org/10.1007/978-3-030-18460-5_14

selection to select the very best orchard clones was acceptable, but for creating breeding population crosses.

The RPBC took over these trials in 2000 and responsibility for the radiata breeding population. The breeding plan for the RPBC was written in 2016 by Paul Jefferson just before his sudden death. He fully understood the importance of backwards selection for seed orchards and forwards selection with cloning for the breeding population.

## 14.1 Main Features of RP Breeding

1953–1967 Thulin (1957) Intensive phenotypic selection, mostly in Central North Island, two clonal seed orchards of 10 ha and 14 clones each.

1972 Assessment and analysis of 1955-planted 29 parent 12-year-old OP progeny tests at 3 sites. Good heritability and genetic correlation estimates of a wide range of traits. Very little family x site interaction.

1970 Assessment of 1957 CP tests with Kaingaroa bulk gave good (first) gain estimates.

1966 Namkoong et al. 1966, Shelbourne 1969, Burdon and Shelbourne 1971. Deterministic gain simulation in planning strategy.

1969 588 '268' phenotypic selections as OP families planted at 3 sites (90,000 trees) not intended as a breeding population: but formed one. There was no provision for AS which was an unknown concept as well as BP, backwards and forwards selection. Schutz and Cockerham sets in reps design were used then and for many years.

1974/1975 Forwards selection of '875's from age 5 '268' OP tests, best 90/588 '268' OP families (too few for AS). 1.5 generation orchard, mating (disconnected diallel) 100 '875' ortets by climbing. Object: backwards selection of '875's for new orchard, not for AS (too much family selection anyway).

1977 IUFRO in Rotorua. Emphasis on mating designs, selection index, but nothing on needs of within-family selection (for AS).

1980 Selection of best '268' families, age 11 years, for backwards roguing of '268' orchards and forwards selecting '880' clones for new orchards. 170 '880' OP families planted in 4 progeny tests.

'Backwards selection' was still driver for '875' and '880's.

1984 CP of best '268's for future AS (the first purpose-built AS families but on too few '268's). Constrained by belief that AS population had to be of very best backwards-selected parents. Combined index selection no good for AS as it involves among-family selection.

1984 First CP orchard at Amberley (near Rangiora, SI) from existing 2-year-old grafts/cuttings. Early, good flowering site. Major change in production population methods.

1985 New theory from Namkoong and Burdon. Multiple populations, sublining, gain pyramid, genetic inter-site correlations.

1985 Shaw and Hood paper within-family (versus among-family) selection for advanced breeding, role of clonal selection.

1986/1987 Development Plan. First recognition of needs for within- versus among-family selection and effects on genetic variation in the population. OP, polycross and NC2 are all good for GCA. Different designs are needed for advanced generation selection (AS). Selection for orchards is quite different from AS selection. Need for cloning for within-family selection. Major rethink about the programme and how to fit the several related clone series into a plan for advanced generation selection.

CP orchard is an important strategy-changing development, with nursery cutting technology. Only two sublines are required.

Breeding strategy still based on backwards selection. Single CP family blocks for within-family selection. OP tests can be used for AS but without among-family selection. Cloning forwards selections in BP will benefit BP and orchards. Proposed Big Bang, two sublines, single family blocks, two trial designs for GCA and AS.

1990 New breeding strategy: CP orchards, five female testers for GCA, backwards-selected BP parents (so nothing much changed) selection for wood properties; multiple populations for the Main BP with two sublines (for CP orchards). 1975 '850' disconnected diallel results mean no regionalisation of BP and no dominance variance: nothing on within-family selection and AS.

1992 Shelbourne compared gains of four different seedling mating designs, seedling and cloned, with cloned full-sib, which gave highest within-family $h^2$ and highest gains (for eucalypts).

1993 Gea stochastic simulation showed higher gains from larger subgroups and also higher gain from more crosses/parent.

1998 Rosvall, Lindgren and Mullin. Stochastic simulation of balanced within-family clonal selection of 24 Norway spruce parents for 10 years showed minor changes in co-ancestry and inbreeding and almost no reduction in additive variance and gain. Best evidence for use of cloning and within-family selection for radiata BP and seed orchard plans.

2000 Jayawickrama and Carson. New elements proposed – the 4 breeds or Elite populations of 12 parents per subline, 24 parents total, based on breeding objectives of growth and form, structural timber, appearance timber and *Dothistroma* resistance, for optimum gain while delaying inbreeding. *But* only GCA-tested clones used for parenting breeds (backwards again). CP orchards are the main seed production method: Elite BPs as two sublines. Same basic plan for large Main population. Controlled crossing in the new breeds. Most of the CP seed is expected to be cloned from young seedlings, as rooting fascicle cuttings was feasible: nothing on AS.

2004 Burdon and Kumar forwards OP gain from orchard at $h^2 = 0.2$ was only a little less than backwards OP, and polycross and pair cross gains were much higher, *but* cycle length of OP short versus CP. Forwards selection with OP families is a viable option.

2005 (published 2007) Shelbourne et al. Deterministic simulation study of balanced within-family selection in Main population from forwards selection, in OP, polycross and pair cross families, seedling and cloned, same number of plants for each family type. Cloned families all gave greater gains than seedlings, up to $h^2$ of 0.5. In these simulations 25-parent Elite used within- and among-family selection of cloned and seedling families which gave much higher gains than in Main BP (with no among-family selection) and higher gains from clones vs. seedlings.

2006 Kumar found good correlation between ramet means in clonal test with the OP seedling offspring of same clones (study done to reassure industry members of RPBC).

2006 Shelbourne et al. Within-family selection: in deterministic comparison of seedlings and clones gave highly superior gain% and gain% per year for clones at $h^2$of 0.2 and 0.5.

2009 (submitted 2007) Dungey et al. (based on report of 2005 meeting in Noosa, Qld).

Jayawickrama and Carson 2000 breeds have been abandoned. Main BP of 500 OP seedling families (250 per subline). Single Elite BP of half seedlings and half clones. Rolling front establishment for single elite. Other Elite populations were abandoned by Dungey et al. But 2018 review abandoned all subdivisions of BP in favour of Kinghorn algorithms.

2016 Jefferson. The Breeding Management Plan. New RPBC breeding programme, begun in 2003, forwards selections from Big Bang (planted 1993) and *Dothistroma* and high wood density Elites (planted 1993–1995), and from '886' and '889' OP progeny test of unselected parentage (planted 1987 and 1990). 360 trees forwards selected. OP seed and scion material collected. OP progeny tests (from 360 selections) in NZ, NSW and Tasmania, 11 sites. BVs of 360 parents, top 1/3 Elite, rest Main population. All 360 clones grafted at Purchase road, Amberley. Crosses used for wide planting. Some CP family seed propagated for clonal tests. Clonal selection to select best clone per family (within-family selection for future breeding population). RPBC breeding population is effectively closed and genetic variation in BP must be conserved.

**I.J. Thulin 1957: Application of tree breeding to New Zealand forestry. NZ Forest Service Tech. Pap. 22 (1957)**

There were reports by Egon Larsen on seed collections in Oregon and California of similar latitude and climate to NZ. Ib discussed provenance problems in Douglas-fir, *P. ponderosa*, *P. contorta*, *P. nigra* and *P. pinaster*.

Larsen noted 50,000 acres of Douglas-fir had been successfully planted in the NI and SI and reported the establishment of the 1957 series of Douglas-fir provenance trials, mainly of commercial seedlots from northern regions of the west coast of the USA. Plans were also outlined for *P. ponderosa*, *P. nigra*, *P. pinaster* and *P. contorta*. By this date of 1957, Ib had a good mental picture of the geographic trends in provenance variation in all these species, based on observations of NZ stands and on early results in trials (1927 Moore study for *P. ponderosa*, published 1944, was the first).

**R.D. Burdon 1986: Clonal forestry and breeding strategies-a perspective. Proc. IUFRO Conf. Williamsburg, Virginia, p. 645.**
CF entails large-scale testing and then planting of clones, selected from CP crosses among intensively selected parents. Without going as far as CF, CP crosses can be produced direct, and vegetatively multiplied. The shift from OP to CP orchards may influence breeding strategies far more than extension to CF. Both need cumulative additive gain from recurrent selection. CP orchards allow much greater specialisation and more flexibility in breed structure. With VM of the seed from the CP orchard, one can reduce lead time. It also allows the use of sublining in multiple populations. CP reduces requirement for female fecundity. Extension to CF can increase breed specialisation and use of maturation, or not, but capability of VM is essential (Burdon 1986).

**P. Cotterill 1986: Breeding strategy: Don't underestimate simplicity. IUFRO Working Parties on Breeding Theory, Progeny Testing and Seed Orchards. Williamsburg, Va. USA. October 13 1986.**
Predicted genetic gains expected per generation and per unit time in both breeding populations and seed orchards are used to compare different strategies for breeding trees. The main selection technique is the combined index, which integrates performance of relatives with individual's performance, in contrast to family and within-family selection. Mating designs examined were disconnected diallel, single pair mating (SPM), polycross and OP.

The combined index is the most efficient selection method, combining family and within-family selection with heritabilities and genetic correlations, whatever mating design was used. Family and then within-family selection is much less efficient. Within-family selection alone is only used in conjunction with SPM [*comment:* but remember the strong family weighting in the combined index which depends among other things on heritability of trait and class of relatives].

Maximum gain per generation is with half diallel. SPM will give the same or better gain per unit time because of long generation times with the diallel. Often a long generation interval with complicated design can make even OP better for gain per unit time.

Basic assumptions by Cotterill were 10 sublines, 30 selections needed for new BP. Research plotted gain *per generation* against $h^2$ and looked at (in descending ranked order of gain) diallel and combined index, SPM and index, polycross and index, diallel and family/within family, OP and index, SPM and within family. Note rapidly increasing gains with increased $h^2$.

Gain per decade is very important. Complicated designs like diallels can get in the way of shortening the cycle. Controlled crossing on ortets can reduce it to 14 years for radiata.

Gains *per decade* ranked are diallel and index (17 years), SPM and index (17 years), polycross and index (17 years) and OP and index (13 years). However,

OP and polycross are almost identical, and both were a rather small amount less than diallel and index, i.e. 4% versus 6%. SPM and index was only 1% behind diallel and index.

***Comment*** An important thing to note is that the index uses family performance, dominantly. This would need a major revamp where there is little family selection planned. However, SPM is a whole lot quicker and cheaper than the diallel, especially with a large population. OP is still the simplest and quickest, but with reduced gains.

Changing to pollination on ortets, all values based on index, brings SPM to the top of the ranking, ahead of diallel and polycross. OP with index is still behind the rest but not by much (4% versus 6%).

This is a valuable paper with ideas I wish I had had available when I wrote the Development Plan. The full version is Cotterill, P.P. 1986. Genetic gains expected from alternative breeding strategies including simple low-cost options. Silvae Genetica 35: 5/6 212–223.

**S.D. Carson 1989: Selecting *Pinus radiata* for resistance to *Dothistroma* needle blight. NZJFS19.**
*Dothistroma* was assessed in 9 progeny tests, at ages 2–9 years. Resistant families could be identified from all assessments, and rankings were constant over years and sites. Heritability was moderate, and SCA was very small. There was no GxE. Greatest net gain is realised by giving as much emphasis to *Dothistroma* as to growth and form.

**R. 2213. C.J.A. Shelbourne, M.J. Carson and M.D. Wilcox 1989: New techniques in the genetic improvement of radiata pine. Commonw. For. Rev. 68 (3), 1989. 13th Commonwealth Forestry Conference Sept. 1989, Rotorua, NZ.**

1. Techniques = strategy, theory and method
2. Hierarchy of populations: the familiar pyramid of seed production, breeding and gene conservation populations. Use of two types of crosses in BP: full-sib SPM or disconnected factorial crosses for BP, and factorial entries crossed as males with five female testers for GCA estimation

***Note*** These are both non-economical designs and probably not the best. The SPM is a better means of advancing the BP (SPM needs only one cross per two parents). NC2 female tester needs five crosses per male tested and is a highly uneconomic means of estimating breeding value or GCA.

3. BP of several special purpose populations, GF, density, *Dothistroma*, multinodal pole, special regional, in ‘Amoeba’, maintained by assortative mating in ‘breeds’ (‘feet’ of amoeba). Only two amoebae are needed with CP orchards. The ‘feet’ changed a lot with future breeding objectives. (Elite concept is introduced here.)
4. Sublining (two at least, amoebae in parallel). Many sublines needed in conventional OP orchard.

5. CP orchards. Many advantages, particularly only two sublines needed for outcrossing in the orchard.
6. 'Backwards' selection on GCA for orchards and combined use of GCA (family mean) and individual performance. 'Forwards' (within-family) selection of new BP (or orchard) parents. Early use of these important distinctions.
7. Previous multisite trials have provided knowledge of GxE to allow restricted planting of BP to 3–5 sites to give stable BVs.
8. Intensive backwards selection has given high gains from '1.5 generation' orchards and for future CP orchards (but not for long-term breeding).
9. Genetic gain trials have compared orchard seedlots and were much needed for gain estimation (versus progeny tests) for justification of programme purposes.

'New techniques' also included vegetative propagation and the new CP orchards. The vegetative propagation advances are radiata specific and so is mass vegetative propagation of young seedlings supporting the hedged CP orchards, now 5 years later into commercial production of expensive seed. New meadow orchards (Sweet and Burdon 1983) are under development. 'Family forestry' is now a reality. 'Field cuttings' and 'juvenile cuttings' were two alternative propagation methods. There was a new commercial-scale tissue culture lab of Tasman Forestry at Te Teko, and much information about progress with clonal forestry; results of genetic gain trials are all radiata pine specific. Reference list was very minimal. Most of the 'news' was unpublished. Randy J. had done a paper in the reference list on 'Value Gains...' in NZ J For. of 1989/1990.

**R. 2367. M.J. Carson, R.D. Burdon, S.D. Carson, A. Firth, C.J.A. Shelbourne and T.G. Vincent 1990: Realising genetic gains in production forests. IUFRO Working Parties on Douglas fir, Lodgepole pine Sitka spruce and Abies spp. Breeding and genetic resources. Session: Genetic Gains in production forests. Tacoma, Wash., USA. August, 1990.**

This is a good paper with 48 references and covers much of the ground of Shelbourne et al. 'New techniques...;.' paper of 1989. It emphasises the growing importance of 'family forestry' and the new CP orchards. There are good diagrams on the history of the radiata breeding programme from 1953, gains from 1960 to 2000, breeding population structure of multiple populations and two sublines, breeding strategy and cycle and breeding generation timetable. An interesting section examines 'possible future developments to 2020'. This paper examines gains achieved by a generation of population improvement and identifies the activities that have led to gain. Breeding strategy is discussed in terms of new features to optimise future gains, including:

- Gains from a generation of tree improvement.
- Species and provenance choice.
- Plus-tree selection in plantations; second attempt led to backwards selection of 94 parents from 588 OP progenies.
- Production population gains: Gains from orchards, backwards selection as above, STP designs raised family mean heritability, needed choice of suitable trial sites.

- Delivery of gains in seed orchards. New CP orchards will produce seed of top GCA clones. Contrast clonal forestry. Reasons for poor gains from conventional orchards were multiple, including external source of pollen, graft incompatibility, insufficient cuttings for orchard planting, too many clones and unsuitable for roguing.
- Development of BP and breeding strategy for *Pinus radiata*

Use of only the best 94 selections ('875's) from the best 50 OP families (out of 588 families of '268's) was too 'narrow' for the future BP. Gain came from only best female parents. '875' selection was too early at age 5 years. '268' CP family 50-tree blocks were seen as best future source of BP selections.

1. New concepts (see Shelbourne et al. 1986–1987). Now based entirely on RS GCA, not SCA. New CP orchards based on best general combiners.
2. Multiple populations and specialised breeds. Growth and form, *Dothistroma* resistant, high wood density, highly multinodal and long internode (completely separate). Crossing within and between breeds.
3. Sublining. Two only in response to CP orchards. Mating allowed within subline, except in orchards.
4. Complementary mating designs. GCA from female tester, field tested in STPs on several sites. Breeding population (called 'concurrent matings') to give 49-tree family blocks (no specifics for AS). According to Shelbourne et al. 1986–1987, these could be disconnected diallel, or factorial but most economical are SPM.
5. Assortative mating. More crosses among top ranked clones. However, Cotterill uses OP and claims better gain per year. NZ uses CP in all BP mating, trying to get better gain per year by overlapping breeding archive and testing by early planting of archives. (There is no acknowledgement of the characteristics of within-family selection versus among-families selection and need to exploit the half additive variance expected within families).
6. Early selection. Early is OK for a few traits like *Dothistroma.*
7. Regionalisation is not necessary, based largely on 11 site 1975-planted '850' disconnected diallel and polycross.
8. Seed and plant production. CP hedged orchards, seed certification and vegetative multiplication.
9. Clonal forestry. Uniformity, also non-conventional traits. Clonal multiplication and storage during long test period are the main problems.
10. Future developments. 'In the future, gains from recurrent selection will depend more on selection within families derived from CP matings of the original parents. Can breeding population gains be maintained or accelerated under this constraint?'

The following biotechnology may be involved in the future:

- Embryogenesis and tissue culture
- Genetically engineered clones
- Gene mapping ID of major genes, gene products and processes
- Tissue culture and maturation

Conventional breeding techniques will involve some new traits, especially wood properties.

**R. 2226. S.D. Carson and M.J. Carson 1989: Breeding for resistance in forest trees-a quantitative genetic approach. Annu. Rev. Phytopathol. 27: 373–395.**
A nice introduction relating plant breeding practice with its inbreds, hybrid varieties and clones to the population improvement of mainly coniferous species, but also eucalypts, Gmelina and Teak. Reliance on recurrent selection and quantitative genetic theory characterise tree breeding for improved disease resistance rather than immunity.

Steps are outlined with a QG approach drawing on experience with *P. taeda* and *P. elliottii* with *Cronartium fusiforme* and *P. radiata* with *Dothistroma pini*:

1. Hazard assessment and integrating selection for *Dothistroma* with other traits. High-risk sites vs. general. Working with populations already selected for other traits
2. Disease resistance and mechanisms
3. Screening techniques
4. Quantitative genetic approach
5. Direct and indirect selection. Strategy and predicted gain
6. GxE interaction and more

A very neat guide to resistance breeding in forest trees as part of an overall breeding and seed production programme should be a text for introducing resistance breeding in any new species, and in AS.

**R. 2382. S.D. Carson 1990: The practical importance of genotype x environment interaction for improvement of radiata pine in New Zealand. 1990 IUFRO Working Parties on Douglas fir, Lodgepole pine, Sitka spruce and Abies spp. Breeding and genetic resources. Session: Genetic Gains in production forests. Tacoma, (Olympia) Wash., USA. August, 1990.**
This was an excellent paper based on the 9-year assessment of 11 NZ sites of the 1975 '850' disconnected diallel. This involved 25 parents and 5 half diallels, 1 full diallel, 50 crosses. Family x site interaction was significant for 6 traits, diameter b.h., straightness, branch habit, malformation, *Dothistroma* and needle retention.

For straightness and branch habit scores, interaction variance was less than half of total variance among crosses. For diameter, GxE was more than half of crosses variance. Genetic gains were predicted for several regionalisation options using a multisite index, but these increased gains over gains-from-all sites only slightly. For diameter, many fewer than 11 sites were needed for selection to capture nearly all the gain from overall selection, nationally.

Some sites were much better than others, and selection on one site gave 90% of possible gain, while another gave only 23% of the total gain. A few well-chosen sites with good growth, high phenotypic variance, high repeatability of family means and high genetic correlations with other sites can do the job.

This was a full report and analysis of these trials, fully documented, more than expected for a conference paper (and a good model for such a study which was so well designed by me in 1970 something!). This paper also carried a good interpretation of the results of such a trial for planning a breeding strategy and regionalisation, or rather the lack of it. This paper is a good treatise on the whole business of GxE and regionalisation.

**G.R. Johnson and R.D. Burdon 1990: Family-site interaction in *Pinus radiata*: implications for progeny testing strategy and regionalised breeding in New Zealand. Silvae Genetica. 39: 55–61 1990.**

An OP progeny test of 170 '880s' was planted at two pumice sites, Taupo and Rotoehu, and two Northland clay sites, Moerewa 1 and 2, both P-retentive. Family x site interaction at age 4.5 years for volume was highly significant between pumice and clay sites but small between two pumice sites. The Northland clay and pumice sites generate marked GxE (from genetic variance and genetic correlation). However, regionalisation only increased volume gain from 22% to 25%. Production of regionalised seedlots does not appear to improve gain; however, progeny testing in only one region results in poor gain for other regions. Progeny tests must be established in more than one region to get a widely adapted tree. There were parent clones whose families did well on a variety of sites but were not necessarily the best. Progeny test site selection is crucial.

**R. 2416. S.D. Carson 1991: Genotype x environment interaction and optimal number of progeny test sites for improving *Pinus radiata* in New Zealand. NZJFS 21(1):32–49.**

This is a repeat of R. 2382 which was a conference paper. This paper sets out to determine if GxE is large enough to warrant a regionalised programme, based on results of a trial of disconnected diallels of 50 full-sib families on 11 sites. The number of tests required to represent a region is a trade-off between costs and benefits. Genetic gains using multisite index selection suggested that regionalisation of seed orchards increases gain only slightly. Choice of site showed that the best site gave 90% of maximum possible gain and the poorest site gave only 23%. 'Good' sites had rapid growth, high phenotypic variance and high repeatability of GCA, i.e. high narrow sense $h^2$.

This is a good general guide to coping with GxE in the other AS breeding programmes.

**R. 2469. T.G. Vincent 2006: Developments in radiata pine seed production. Proc. 11th RWG1 (Forest Genetics) Meeting, Coonawarra, SA, 1991**

*A seed production system needs:*

1. Ability to produce large quantities of seed with desired levels of genetic gain, quickly
2. Full control of contributions from male and female parents
3. Flexibility in breed type produced
4. Cost competitiveness with other plant production systems

*Why use CP orchard seed?*

*Traditional seed orchards:*

1. Advantages and limitations of inflexibility in breed production
2. Low productivity per ha. of orchard
3. Need for minimum number of clones in OP orchard
4. Inability to control pollen cloud
5. Rapidly increasing height above ground of seed production over years

*Control-pollinated orchards. Advantages are*:

1. Control of both parents means increased gain and possibility of diverse breeds.
2. Smaller tree size.
3. Earlier and heavier seed production.
4. Close spacing means better land utilisation.
5. (A)dvanced (S)elections can become available quickly.
6. Best cone producers used for seed production.
7. Cones collected green and a year earlier.
8. Specific crosses easily done.

Management of CP orchards: Several management factors include 'good' siting for seed production and adjustment of hedging and hormone treatments to each clone.

Seed production capability of a CP orchard is limited. Main problem is in development of cost-effective vegetative multiplication of CP seedlings.

**R.D. Burdon, M.H. Bannister and C.B. Low 1992: Genetic survey of *Pinus radiata*. Introduction, description of experiment and basic methodology. NZJFS 22**

This is the introductory paper in a series of nine which cover all aspects of the initiation and subsequent studies of this very large provenance-OP progeny study, located on two sites in Kaingaroa. Rowland's and Martin's 'life work!' Covered in nine successive papers in NZJFS 22.

**C.J.A. Shelbourne 1992: Genetic gains from different kinds of breeding populations and seed or plant production populations. S. Afr. For. J. 160:49–65.**

This was originally a conference paper for the IUFRO Symposium "Intensive forestry: the role of eucalypts" Durban Sept. 1991 and was a paper about deterministic gain prediction, using the same basic equations of Namkoong et al. 1966. The scenarios were of different OP, CP and cloned breeding populations and clonal and seedling orchards under contrasting heritabilities of 0.1, 0.2 and 0.4. Gains from clonal forestry were also simulated. Cycle length and costs for different options were examined for all scenarios. Five breeding population options were covered, from solely phenotypic selection, BP of OP families from the wild, ditto from a clonal archive, BP of full-sib families to a cloned full-sib breeding population.

Twelve different production population options were covered. It systematically looked at the possible options, suited to eucalypts.

The gains were shown from a breeding strategy (that was most used in our 'alternative species') of an OP breeding population with OP seed collected from select trees in NZ plantations of possible local land race, or from native populations in country of origin, with forwards selection from these progenies. Because the new forwards selections were made in a thinned progeny trial, I predicted gain from the female (selected) parent plus the gain from the male parent (which benefited from the thinning in the test). This made the gain prediction more complicated but roughly equivalent to selection from only the best 2/3 of the families (67/100) and the best tree within those families of either 40 or 100 trees each. The gains from, e.g. $h^2$ 0.2 with such a low among-family selection ratio, were slightly less than for simple phenotypic mass selection. Using full-sib CP families with the same among- and within-family ratios only raised the gain from 8.4% to 9.1%. However, if the BP families were cloned, with 10 clones per family with 10 ramets per clone, the gain (in the $h^2$ 0.2 selection) increased strongly to 14.1%.

It was clear from this and other analyses that even low intensity among-family selection in the BP, necessary to preserve status number (effective population size), was not going to give big gains each generation. Only by cloning could one get some 'high heritability' selection done within family, to boost BP gain, without reducing status number.

In the production population options, a clonal orchard from intensive among-family and within-family selection gave a gain of 15.5% ($h^2 = 0.2$), and a further roguing, keeping only 1/5 of the clones on later test results, boosted gain to 20.7%.

Clonal selection in a clonal forestry scenario from OP families in the original OP test increased the gain dramatically, gains depending a lot on how many ramets per clone were used but would require ongoing clonal testing and then repropagation of best clones, taking another test cycle.

These gain figures for the three $h^2$ of 0.1. 0.2 and 0.4 gave a clear picture of the effects of trait heritability. Deterministic simulation of this kind is only for a single generation but is sufficient to show what happens to gain in various BP testing and orchard situations. It does bring out the importance of cycle length for different scenarios. It was further attempted to translate the gain per year into gain per dollar (Shelbourne 1992).

***Comment*** These results could be applied in the NZ context, both for radiata and the eucalypts and Douglas-fir, though none of the alternative species was cloning the BP yet feasible. The cloned BP concept has been later adopted for radiata by the RPBC and also apparently in the NC State Cooperative for P. taeda, promoted by Fikret Isik (now associate director of their NC State-Industry Coop.).

**R. 2992 R.D. Burdon 1995: Future directions in tree breeding: some questions of what we should seek, and how to manage the genetic resource. CTIA/WFGA meeeting, Victoria, BC August 1995.**

Gains in productivity in crop breeding, i.e. in harvest index and grain yield, are quite different from gain in tree volume per ha. For trees, the breeder needs to look at exploiting divergences between crop productivity and competitive ability. There are several problems. Monoclonal blocks? How to translate gain from single-tree data to final whole crop production? Economic weights and effect of density plus volume? Gains in pioneer species vs climax species?

With geographic transfers in native species and in exotics, local indigenous material may be suboptimal because of lags in adaptation and limited genetic bases. As one example, contrast movement of Douglas-fir north-south (not sensitive) with movement of Sitka spruce in relation to weevil.

Wood properties: high $h^2$ but difficult and expensive to measure.

In management of genetic material, the role of conservation of gene resources is fundamental, and a large BP is needed as well. Inbreeding leads to natural elimination of inbreeds, but tolerance of levels of inbreeding is unknown. This has a role in sublining and subline size, and the effect of very small breeding groups, as in the new NC State program.

- There is a need for underpinning breeding by gene resources.
- The role of biotechnology in gene resource conservation has been 'not much'.
- Long exposition on how to handle gene resource populations, from NZ experience with radiata.

In the situation of a cooperative breeding program, severe strains can arise from rapid advances in biotechnology, as there are different levels and commitment to biotechnology in different companies. Proprietary appropriation of technology may lead to a defensive spiral and detriment to scientific communication (Burdon 1995).

***Comment*** This has some good angles, but it is a long-winded discourse, not easily summarised, which tends to bury good points. Position in breeding of native species, especially in the PNW, is much trickier than with exotics. Important take-home lesson for AS is that the use of good provenances is very critical, especially for growth gain.

**D. Lindgren, L.D. Gea and P.A. Jefferson 1997: Loss of genetic diversity monitored by status number. Silvae Genetica 45 (1):52–59.**

Monitoring genetic diversity (effective number) in a BP can involve a new concept, status number = ½ 1/ average co-ancestry. It can be preserved over generations by repeated selfing (!). Using equal parental representation, diversity is preserved better than by panmixis. Small sublines preserved diversity better than large ones but only when sublines were very small (like repeated full-sib mating). Rapid decline in status number seems unavoidable.

**P.A. Jefferson and S. Weaver 1997: Gains from clonal testing in the breeding population. IUFRO '97 Genetics of Radiata pine. Proc. NZFRI_IUFRO Conference. Dec. 1997, ed. R.D. Burdon and J.M. Moore. FRI Bulletin no. 203.**
Gains were calculated from cloning a breeding population of 150 trees in single pair mating with 2 individuals selected per family from a base population of 1000. Three levels of $h^2$ were 0.7, 0.3 and 0.1. Additional clone-within-family effects (as experienced in some clonal tests) were assumed at 10, 30 and 50% of additive variance. Gains were compared with fixed and unlimited resources. 36 seedlings per family with fixed resources gave highest gain at $h^2$ of 0.7. Clonal or seedling gains were best at $h^2$ of 0.3 and 0.1.

Increased resources gave increased gains. Cloning can increase gains in traits of low $h^2$. This stochastic simulation looked at gains from family and within-family selection at three $h^2$s. It did not specifically look at gains from within-family selection.

**R. 2622. R.D. Burdon, A. Firth, C. B. Low and M.A. Miller 1997: Native provenances of *Pinus radiata* in New Zealand: performance and potential. New Zealand Forestry 41(4): 32–36. (Ken Eldridge should have been included as a coauthor).**
This paper is mainly about the results of the 1978 Firth-Eldridge collection of seed from the three Californian provenances and from Guadalupe and Cedros islands and is designed for a general forestry audience, not a NZJFS audience. There were 4 subpopulations at Ano Nuevo, 6 at Monterey and 3 at Cambria, and 22 to 70 parent trees of each subpopulation were climbed for seed in each locality. An additional three bulk lots were obtained from Kaingaroa, Nelson and Southland (NZ lots were of somewhat select seed trees in the respective forests). All these seedlots were raised in replicated nursery layouts at FRI and Rangiora.

Trials were planted at 23 sites throughout the country from Aupouri to Longwood forests. Design was 12 reps of 6-tree row-plots at 17 sites and 6 main trials of 10 reps of 36 tree plots; 5 reps only were planted at Berwick and Longwood. Assessments were timed when heights were between 7 and 10 m from age 5.5 years onwards, depending on growth. Traits were DBH, bark thickness, predominant height, survival, straightness, branch cluster frequency (1–9 scores) and stem acceptability. *Dothistroma* and *Cyclaneusma* needle-cast were scored when expressed.

Ano Nuevo and Monterey are best adapted overall to NZ conditions. Ano Nuevo is less well adapted to P-deficient clays (North of Auckland) and better adapted to cold and snow sites. (Monterey was best adapted to warm and infertile clay soils.) Cambria is susceptible to needle-casts, shoot dieback and frost and snow damage but has tolerated poor soils. Guadalupe, with adaptational problems, grows more slowly than the mainland provenances but has very straight stems, 10% higher wood density and grows well in hybrid combination with Ano Nuevo and Monterey. Local NZ lots are 50–75% Ano Nuevo and the rest Monterey (and outperform native lots).

In the report, a Table of Relative Performance for DBH and Basal Area (of the large plot trials only) of the four main provenance groups, Ano Nuevo, Monterey, Cambria and NZ, shows clearly that Monterey was superior to Ano Nuevo on all sites except southern South Island, yet not to the NZ land race populations. However, this result applies only to the early 7–10 m height data in the paper. These results confirm the wisdom of using NZ land race stocks for breeding programmes.

This was a major provenance testing operation on our main species which (amazingly) had never previously been properly performed. (A very limited 1955 experiment and an extension of the Genetic Survey of some surplus stock on four sites made in 1968 were the only 'national' provenance trials of PR). The two Genetic Survey experiments in Kaingaroa gave a detailed picture of provenance and genetic variation, but genetic parameters of native populations were not entirely representative of the NZ land race in which breeding was taking place. The gene conservation planting of bulked subpopulations were planted in various parts of NZ, mostly Kaingaroa, which are important reservoirs of genetic variation, as are the trials themselves.

The future of these trials and the related conservation areas has been in some doubt. We are lucky to have secured these materials since movement out of California has been constrained by pitch canker. This whole operation has many lessons for AS which have been applied in the Low-Miller collection of coastal Douglas-fir made in 1993. These actions taken in what seems a Golden Age of tree breeding need our respect for the dedication by the people who did them, and for the wisdom shown by those who instigated and funded them.

**K.J.S. Jayawickrama, M.J. Carson, P.A. Jefferson and A. Firth 1997: Development of the New Zealand radiata pine breeding population. Proc. IUFRO'97 GENETICS OF RADIATA PINE, Rotorua, NZ 1–4 Dec. 1997 (FRI Bulletin No. 203).**

Radiata pine breeding work from 1953 to 1997 is reviewed, with most emphasis after 1985 (Development Plan was 1986). During the 44 years of the programme, 2650 trees have been progeny tested, and 1500 crosses were made for AS. It was noted that (1) OP tests for within-family selection were replaced by CP. Underlying approach was recurrent selection for GCA. (2) CP orchards are now the main seed production method; (3) BP is now divided into two large sublines; (4) major crossing programme (disconnected factorials with 50-tree blocks of full-sibs) in late 1980 s generates candidates for AS. (5) Parallel testing for GCA of clones of different selection series is now using NC2 with five female testers.

Work is in hand on developing additional breeds (small Elite BPs) in addition to existing 1970 Long Internode Elite. This includes 2 sublines per breed/elite and 12 genotypes/subline/breed. Threshold values for growth rate, stem form, *Dothistroma*, *Cyclaneusma*, stiffness and spiral grain for all breeds/elites are maintained. Proposed new breeds are *long clears*, *structural*, *appearance grade*, *fertile sites* and *Dothistroma resistant*.

This description is for the elites/breeds, small 24 parent populations. A similar but less intense programme is being developed for the large Main population. (It should be recognised that RS/GCA is a backwards re-selection strategy; candidates for each Elite are the product of crossing of clones whose GCA has been determined by a previous generation of polycross, OP or NC2.)

Reasons for the initiation of breeds (elites) include greater genetic gain in product-related traits, convenience, crossing smaller number of genotypes and more precise estimates of GCA.

The rest of the long paper is a history of the development of the breeding programme starting with the RPBC (1987–1997), CP orchards, land race selections, advanced generation selections (mostly in CP trials but often at age 4, needing more testing), Australian selections and selections in native provenances.

Systematic progeny testing with OP was used for 1969-planted '268's onwards, later use of polycross for 100 '850's and 50 '850', in disconnected diallels, for regional GE and genetic parameters. Regionalisation was abandoned after Sue Carson's analysis of 1975 disconnected diallel indicated there was no need. After the 1986 Development Plan, CP orchards became the accepted future strategy, accompanied by two sublines only.

Nearly 1000 selections were made in 1986–1988 in unimproved 'land race' stands as a last attempt to bring in selected genotypes unrelated to the existing breeding population. A large number of selections were made in progeny trials from 1986 onwards. Selections were made at 8+ years. 270+ selections were made in native provenances, 1964–1967, 1978 and 1980. Mostly '894' series, 80 of these were Guadalupe.

Use of female tester design was partly to standardise the criteria of selection. The NC2 with 5 female testers and 206 parents crossed gave 958 crosses. (*Comment.* This in my view was crazy. 200 polycrosses would have done the job.)

The highest ranked selections were intercrossed for future advanced generation selection, incorporating strong among-family selection. However, GCA ranking from female testers was separated from AS selection. Also, now there were two superlines. The Big Bang was done with 3-2-1 representation of crosses but partly failed because of caterpillar defoliation. (*Comment.* This approach to AS seems totally at odds with any attempt at minimising loss of diversity and inbreeding and maintaining status number.)

Controlled crossing is planned in the new breeds: long internode, high wood density, *Dothistroma* resistant and GF. Most of these crosses will be cloned. By 1997, rooting of fascicle cuttings was feasible, and breeders thought that clonal replication would be effective for within-family selection. Reduction in size of BP down to 400 will make it easier to find the best genotypes. Elites will be targeted at specific sites, along with their cloning. (*Comment.* No consideration about amount of among-family and within-family selection.)

With the further developments of breeds, there is concern about shorter rotations, wood quality and form on ex-pasture sites, but increased gain from the breeds will target specific products. Breeds will link BP and seed orchards. Smaller breed BPs will derive from CP seed from 2 sublines, 12 parents per subline. Other prescriptions for breeds include 6-parent disconnected factorials, seedling progeny cloned for testing at 4 locations, one site required with plenty of flowers and catkins to create next generation. Elites will be targeted at specific site types and the use of clones confined to those sites. (*Comment.* Thus, more within-family selection [YES!], but 6-parent disconnected factorials were not needed, SP crosses would do.)

**S.E. McKeand and F.W. Bridgwater 1998: A strategy for the third breeding cycle of loblolly pine in the southeastern United States. Silvae Genetica 47: 223–234.**
This strategy will be managed in a hierarchy of three populations per region, each with different levels of selective breeding. The mainline population will consist of 160 selections, available to each cooperator *in a given region*. These populations will be managed in a sublined BP of 40 sublines of 4 trees each to provide long-term genetic gain and diversity.

The most intensively selected and managed hierarchy will be the Elite population. This highly select group of about 40 will be managed to provide short-term gain.

A third hierarchy will be a genetic diversity archive, preserving genotypes with extreme BVs for certain traits, as an insurance population.

Improved efficiency of this strategy (reduced population size) means reduced effort and cost. There will be increased selection intensity and reduced time for a cycle because of reduced number of crosses. This will mean better gain per decade. More effort will go to the elite.

**G.R. Johnson and J.N. King 1998: Analysis of diallel mating designs. I Practical analysis procedure for ANOVA approximation. Silvae Genetica 47 (2–3):74–79.**
Computer programme for analysis of diallel.

**J.N. King, M.J. Carson and G.R. Johnson 1998: Analysis of disconnected diallel mating designs II. Results from a third-generation progeny test of the New Zealand radiata pine breeding programme. Silvae Genetica 47: 80–87 (1998).**
Genetic parameters from a second-generation disconnected diallel had $h^2$ of 0.2 for DBH. Coefficients of variance for additive and phenotypic variance were the same as the previous generation. There was a trend of declining dominance variance and increased additive with age up to 7 years.

*Conclusion*: Progenies from two generations of selection exhibit larger amounts of additive variance, like those estimated in earlier generation progeny tests, i.e. $h^2$ 0.2 for growth and yield. This confirms strategy of population improvement using recurrent selection for GCA. Second conclusion is that dominance genetic variance for growth traits tends to disappear with time (by age 7). Selection for best GCAers should be interpreted cautiously.

**K.J.S. Jayawickrama and P.A. Jefferson 1999: Stochastic simulation of the genetic advancement of multiple traits using sequential culling. NZ J. For. Sci. 29: 25–43.**
A software tool was developed for stochastic simulation of multiple traits and used sequential culling for the different traits. It is applied to the BP for specific products and sites. Gains at age 8 years from different selection emphases are compared for an unspecialised Main of 300 parents and 4 breeds of 24 parents. Forwards selection is covered in progeny of selected parents. Breeds were structural timber, GF Elite,

long internode and fibre, reflecting different end-product objectives. The simulation supported differentiating a structural timber breed with strongly multinodal branching, high wood density and reduced spiral grain as well as similar breeds to existing LI and GF.

**K.J.S. Jayawickrama and C.B. Low 1999:** ***Pinus radiata*** **selections from different regions in NZ differ in branch habit, form and growth rate. NZ J. For. Sci. 29 (1) 3–21**

Data at age 8 years were compared from two large trials, each of plus-tree progenies of selections from different regions in NZ. These were the '850' polycross of 109 parents at 6 locations and the '888' test of 239 OP progenies from trees selected in 6 regions and from a Chilean orchard. In the '888' trial, there were significant differences for every trait at all sites except needle retention. Southland and Nelson sources had low branch cluster frequency scores. Branch frequency was the only trait with significant differences in the '850' polycross.

The conclusions were that there was faster growth and better form in second rotation stands in the Central North Island and lower branch cluster frequency scores in the South Island, possibly because of better ability to select.

**K.J.S. Jayawickrama and M.J. Carson 2000: A breeding strategy for the New ZealanRadiata Pine Breeding Cooperative. Silvae Genetica 49, 2.**

This paper documents the proposed new strategy for breeding PR in continuation of the 1997 paper on 'development of the New Zealand breeding population'. Emphasis is on RS/GCA (i.e. additive genetic), but it should be noted that this originally meant reselection for GCA, or backwards selection, as in roguing or creating a 1.5 generation orchard.

The strategy is based on additive genetic effects (RSGCA), and a two superline structures for Main population and breeds (Elites). A non-regionalised BP is maintained, with final selection at about age 8 years.

New breeds (Elites) are *structural timber, clear cuttings, growth and form* and *Dothistroma resistant*. Existing LI and high wood density breeds are to be used as sources of selections. Combined population census number for breeds is 4 × 24 = 96. Census number of total population is 550 and target status no. is 400. (Note this does not match with Paul Jefferson's estimates of 29 and 33?)

The role of breeds is to get optimum gain while delaying inbreeding. The Main population is a reservoir of genetic diversity. Incidentally, the 1986 strategy for the Development Plan had RS/GCA, two superlines, unspecialised Main and specialised breeds/elites, separate crossing for GCA and recombination (AS), and these aspects will be maintained. 2650 clones have been progeny tested, and 1500 clones have been CP tested with 1300 in archives.

The base populations are the five native populations. NZ land races have generally outperformed these.

'Breeding objectives' are more easily determined than 'breeding goals' of which there are 4: growth and form, structural timber, clear cuttings, *Dothistroma* resistance.

The previous 'long internode' feeds 'clear cuttings', and 'high wood density' feeds 'structural'.

The introduction provides a complete description of the background to this strategy, with all relevant references. The previous strategy is well described in terms of selection traits, superlines, non-regionalised programme, structure of Main and breeds populations. Selection traits are DBH, straightness score, branch cluster frequency score, needle retention score, malformation score and acceptability.

There are two superlines, to match CP orchards.
The breeding programme is not regionalised.
The breeding population is partitioned into a Main and several (at least 4) breeds.
The 'breeds' are now discrete control-pollinated populations.

In selection strategy RS/GCA (recurrent selection for additive effects) has continued to be emphasised, referring to Namkoong et al. (1966) and Shelbourne (1969). RS/GCA is a method of re-selection, targeted at a clonal seed orchard using an NC II, polycross or clonal test for roguing or reselection. The Namkoong 'control pollinated seedling seed orchard' represents a single generation of crossing with between-family and within-family selection, in fact AS. This was directed at seed orchards and not directly at breeding strategy.

OP, polycross and later NCII female tester were recommended entirely for GCA estimation. GCA ranking and recombination (AS) were therefore separated, though no explanation was given for the weighting given to among-family selection.

Formulation of economic weights is a continuing problem and case for research.

## 14.2 Revised NZRPC Strategy

The key features are the two superline structure, emphasis on recurrent selection for GCA (additive effects), non-regionalised breeding programme and final selection at age 8 years in CP tests. Priority is for breeds versus Main and Guadalupe populations. High priority is given to improving corewood. Only GCA-tested clones are used in current cycle.

Two new breeds are to be formed (clear cuttings and structural timber). Breeds will be tested as full-sib families, either as seedlings or clones within families. Existing breeds' growth and form and *Dothistroma* are unchanged.

Many GCA-tested clones exist so that only clones which have been GCA tested are used. The 4 breeds, growth and form, *Dothistroma* resistant, clear cuttings and structural timber, all 24-parent, make a lot of sense. I support the breeds (Elites) concept and Figs. 14.1 and 14.2, ('A breeding strategy for the New ZealanRadiata Pine Breeding Cooperative') especially 2 which clearly diagrams the strategy. The Main is an excellent foundation of genetic variability and essential for the security of the narrow-based breeds.

Cloning 10–20 seedlings from each breed family is planned. These presumably are from control crosses for AS, to generate the next generation of the breed.

**Fig. 14.1** Within-family selection in single-family 50-tree blocks. Note: best tree in that family

**Fig. 14.2** Genetic gain trial, FRI Long Mile. Coop field meeting

Within-family selection should be much more effective with clones-in-families. 'Cotterill 1986' needs consulting about the gains expected of the various crossing designs, especially SPM on ortet versus disconnected factorial and diallel in archive, though his use of an index introduces an among-family element in the selection. Gain per year will outweigh gain per generation in the future.

**C.J.A. Shelbourne 2000: A NEW strategy for radiata pine??**

I was lying sleepless last night and thinking of a way to advance a PR generation. This is the simplistic result.

First select parents and collect their OP seed. Establish OP tests on 4 sites and one clonal test (15 clones per family, 10 ramets each) from the same OP families. Restrict clonal test to one fast-growing and good-flowering site (the clonal test is primarily for within-family selection which may be done on one site).

Carry out intensive among-family selection for GCA in the OP tests after 8 years. Use these results in conjunction with clonal test within-family selection for selecting seed orchard clones.

At the same time, do clonal selection within-family in the single clonal test and collect OP seed from the best clone in every family in the clonal test for AS (to establish future BP). Rejection of families for the future BP should be very limited, and gain relies mainly on clone within-family selection.

The production population (CP orchard) relies on selection from results of the GCA tests and on clone within-family selection. Generation time is minimal (only OP, no archives or crossing). Note: the clonal test is a critical element and needs plenty of ramets per clone and intensive management. It can be converted to a seed orchard and is a source of OP seed of the selected AS clones. Move on to the next generation, quickly, via OP seed. The emphasis throughout should be on 'gain per year'.

**D. Danusevicius and D. Lindgren 2002: Efficiency of selection based on phenotype, clone and progeny testing in long term breeding. Silvae Genetica 51:19–26.**

The objective is maximum annual progress in GMG/Y, which is the weighted average of breeding value and gene diversity. Highest GMG/Y was based on 0.25 (clone), 0.152 (phenotype) and 0.139 (progeny). Clone was best except at highest $h^2$. Phenotype is a second best strategy depending on reproductive maturity (note: this utilises family selection, phenotypic selection and clonal selection, and is not targeted at within-family selection).

**M.B. Powell, T.A. McRae, H.X. Wu, G.W. Dutkowski and D.J. Pillbeam 2004: Breeding strategy for *Pinus radiata* in Australia. IUFRO Genetic Proc. Charleston, South Carolina.**

The new strategy is a radical revision of that from the early 1990s (and White et al. 1999) and integrates deployment with the breeding program. More effort is directed at delivery of genetic gain per unit time. Generation interval is reduced and selection pressure increased. Breeding objectives will in the future be defined and developed using economic information, including DBH, straightness, branch quality, *Phytopthera*, *Dothistroma*, density and wood quality.

BLUP is the preferred method for predicting BVs. For rolling front breeding and selection, TREEPLAN uses BLUP. No more discrete generation breeding. Use rolling deployment with overlapping generations. No discrete Main. Parents are used in several crosses (3–4) and no polycrosses. Sublining is to be continued. Size of BP is reduced. Multiple subpopulations of 30–40 will target different BOs. Separate deployment breeds.

Importance of cloning in progeny testing is being reviewed (and for deployment): more crosses per cycle, greater focus on forwards selection for both breeding and deployment and more focus on pedigree control with FS crosses.

**S. Kumar 2006: Correlation between clonal means and seedling progeny means, and its implications for radiata pine breeding strategy. Canadian Journal of Forest Research 36: 1968–1975.**
OP offspring of individual genotypes is commonly used for estimating their BVs or GCA. Alternatively, vegetative propagules could be used to evaluate relative performance. The mean of vegetative propagules may or may not relate to GCA. The main objective of this study was to correlate the performance of clones, propagated by fascicle cuttings in a clonal test, with performance of their OP offspring in a progeny test.

Empirical correlations between mean of ramets and mean of progeny were 0.56 (DBH), 0.63 (straightness), 0.81 (branch cluster number) and 0.09 (malformation) for DBH, straightness score, branch cluster number and malformation score. Theoretical expectations of these were 0.55, 0.69, 0.81 and 0.24. Results of this study indicate that clonal performance of selections in Main and Elite BPs could be used for direct selection in seed orchards with no need to do progeny testing or forwards selection.

**39460 C.J.A. Shelbourne, R.D. Burdon, H.S. Dungey, L.D. Gea and S. Kumar 2006: Revisiting the elite. RPBC/Ensis report? Brain-storming workshop 28th February 2006.**
At the Noosa workshop in March 2005 (Dungey et al. 2006a), several proposals were made for a future NZ radiata breeding strategy about which there was some disagreement by other NZ tree breeders. At the subsequent workshop in February 2006, reported here, deterministically simulated gains of some Noosa plans were examined, involving seedling families and clones derived from these, and where necessary were corrected.

There were also several unresolved problems with the Noosa proposals:

- Several small Elite BPs (breeds) were proposed by Jayawickrama and Carson 2000, namely, structural timber, appearance timber, *Dothistroma* resistance and growth and form, but only a single undifferentiated Elite was proposed at Noosa.
- Half adoption of cloning and half seedling populations in the Elite will be likely to result in major difficulties in interpreting breeding values of seedlings versus clones.

- Selecting high heritability traits with seedlings versus low heritability traits with clones is going to be confusing and may not work, as clones will give good resolution of genotypic differences up to $h^2$ 0.5.

The Rolling Front concept was developed for a eucalypt whose flowering was erratic and unpredictable, delaying breeding work, which is not the same situation for PR, and a Rolling Front approach is unnecessary.

The gain simulations below were unfortunately ‘scrambled’, and the corrected results of the simulations are shown below.

*Noosa Option 1*

Seedlings: 125 full-sib families from 50 parents, 100 seedlings/family 12,500 plants
Clones: 50 families (of the same seedling 125 families), 25 clones per family, 10 ramets/clone 12,500 plants

*Gain %, Gain % per Year and Gain % per Year per 1000 Plants*

Seedlings: 10.8, 0.72, 0.058 $h^2$ 0.2
Clones: 13.2, 1.10, 0.088 $h^2$ 0.2
Seedlings: 13.3, 0.89, 0.071 $h^2$ 0.5
Clones: 17.9, 1.49, 0.119 $h^2$ 0.5

Gain % per year and gain % per year per 1000 plants are much higher for clones than seedlings for both heritabilities. The different length of breeding cycle, 15 years for seedlings versus 12 for clones, is also having a large effect on gain% per year.

Gains from three additional options at $h^2$ of 0.2 were simulated for clones only, involving fewer parents and fewer full-sib families but with the same number of 25 clones per family and 10 ramets per clone. Gains were only slightly reduced with 12 parents versus 20 parents, or 24 families versus 50.

Option 1a: 20, 50, 12,500 Gain % 17.7, Gain %/year 1.48, Gain %/year/ 1000 plants 0.118
Option 2a: 16, 32, 8000 Gain % 16.5, Gain %/year 1.38, Gain %/year/ 1000 plants 0.171
Option 3a: 12, 24, 6000 Gain % 16.4, Gain %/year 1.37, Gain %/year/ 1000 plants 0.228

**C.J.A. Shelbourne, S. Kumar, R.D. Burdon, L.D. Gea and H.S. Dungey 2007: Deterministic simulation of gains for seedling and cloned main and elite breeding populations of *Pinus radiata* and implications for strategy. Silvae Genetica 56: 253–300.**
Deterministic equations for predicting genetic gain were first published by Namkoong, Stonecypher and Snyder in 1966, and these form the basis here of predicting genetic gain from one generation of selection in large Main and small Elite breeding populations for a range of heritabilities. Varying numbers of parents, families, seedlings/family, clones/family and ramets/clone were input to aid revision of the New Zealand *Pinus radiata* breeding strategy. Two sublines would be required for deployment through CP seed orchards.

– Cloned versus seedling populations of equal numbers of plants were simulated, derived from forwards selection in open-pollinated, polycrossed and full-sib pair-crossed families. *Only balanced within-family selection* was used in the 200, 600 and 800 parent Main breeding populations, but among- and within-family selection was used in the 25-parent Elite populations.

Predicted gain from within-family selection in the Main population was highest from the cloned polycross at all heritabilities and lowest for seedling full-sib families. Gains from these populations when cloned were all higher than seedling equivalents at heritabilities <0.5.

Elite populations of seedlings from 25 parents with intense within- and between-family selection resulted in much higher gains, and these were highest from the cloned options.

A new breeding strategy for NZ's *P. radiata* was proposed based on the simulation results, involving forwards selection within families in a large Main uncloned open-pollinated breeding population, supported by parentage reconstruction using DNA markers (the Main was not cloned because of lack of experience with this technology). Forwards selection within and between families in small cloned pair-crossed Elite populations was proposed as the main source of clones for OP and CP seed orchards.

Re-establishing the Main from OP seed for the next generation would be straightforward and involve a minimum time over and above the 8-year growth period. Selection and breeding the Elites would either involve establishing breeding archives or doing controlled pollination on the selected ortets; single pair mating would be the most economical (see Cotterill 1986).

**R.D. Burdon 2008 Breeding radiata pine-historic overview. NZ J. For. 52: 4–4.**

This is a nice overview of the progress of radiata improvement from 1951 onwards. It outlines the breeding goals and options of long versus short internodes and includes the breeds of Jayawickrama and Carson, defining wood properties, and use of mass multiplication of seedlots through CP orchards. Not a key report that needs quoting.

**R.D. Burdon 2008 Branching habit in radiata pine-breeding goals revisited. NZ J. For. 52 (4):20–23.**

Past breeding has pursued short and long internodes as indirect selection for branch size, but it doesn't allow long clears or effectively control branch size. The long internode type has more drawbacks. An alternative ideotype has two coequal branch clusters over a considerable length of stem. Clonal forestry is probably necessary to capture this.

**H.S. Dungey, J.T. Brawner, F. Burger, M. Henson, P.A. Jefferson and A.C. Matheson 2009: A new breeding strategy for *Pinus radiata* in New Zealand and New South Wales. Silvae Genetica 58, 1–2 (2009).**

The strategy presented here is the result of a review by several experts in 2005. This report outlines the preceding strategy of Jayawickrama and Carson 2000, entitled

'A breeding strategy for the New Zealand Radiata Pine Breeding Cooperative'. This report also summarises two lines of research, commissioned by the RPBC. One by Shelbourne et al. (2007) (provided to the RPBC in 2005) was on deterministic comparisons of gain from seedling and cloned Main populations (with balanced within-family selection) of OP, polycross and full-sib families under a range of heritabilities. Gains of clones versus seedlings were compared, and clones had higher gains up to heritability 0.5–0.6. A new strategy and plan was proposed for developing a balanced breeding population of OP families. Seedling families only were proposed for the Main because of inexperience of the cloning technology. A fully cloned EP based on intense between- and within-family selection with 2 to 2.5 families per parent was proposed with at least one clonal test converted to a crossing archive and CP orchard. The gains from clones were always higher than from seedlings up to heritability 0.5–0.6, and the proposed strategy was used as a basis for the strategy review process.

The study by Kumar (2006) compared the breeding values of clones with their OP offspring over 20 years later. The clones were propagated by very juvenile fascicle cuttings. The higher the heritability of the trait, the higher the correlation of clonal and progeny BVs. Clones appeared to be adequate predictors of seedling BVs, so it would be unnecessary to progeny test clonal selections.

In July 2005, a workshop was held to develop a new breeding strategy for the RPBC. The review was based on the Jayawickrama and Carson strategy and on the results of the two research studies outlined above. The proposed strategy is therefore based on a consensus of the participants in the workshop. This was followed by another workshop (Shelbourne, Burdon, Dungey, Gea and Kumar. *Revisiting the elite*, February 2006).

The new strategy comprises a large open-pollinated (OP) Main population (MP) with 500 female parents and 2 sublines (250 parents per subline). The MP will be managed in discrete generations, with selection using 'forwards' selection in OP families as well as some 'backwards' selection using previously GCA-tested parents. Inbreeding will be controlled by parental reconstruction and/or estimating group co-ancestry. Any inbreeding is contained within each subline and negated in the CP orchards. The OP strategy is relatively inexpensive and, without grafting and crossing, will thus reduce the length of the breeding cycle. The Main's OP trials can be used for both GCA testing and forwards selection of the future OP breeding population. It was assumed that parental reconstruction would be further advanced in future.

Rosvall et al. (1998) in a stochastic simulation with balanced within-family selection in a cloned population, double pair mating and only balanced within-family selection over 10 generation, noted that provided each population contained at least 24 individuals, increases in group co-ancestry and inbreeding had no effect on genetic gain, though status number did reduce from 48 to 8. The NZ Main population will be tested as OP families in 8 field trials, 4 for each subline. The level of GxE is still in some doubt, following Carson (1991) who concluded GxE was not big enough with 10 sites to warrant a regionalised breeding strategy. 250 OP families will be included in each subline with 20 seedlings per family per site. The field design will be an incomplete block design with single-tree plots.

There will be a single small control-pollinated Elite population (EP), planted half as seedlings and half as clones. The offspring of 24 parents will be tested each year as clones and offspring of a further 24 parents as seedlings (generating the Elite BP will take place every 2 years, following a Rolling Front establishment plan. The two sublines of the Elite will be maintained as in the Main (the number of crosses per parent have not been shown).

The 'clones' referred to are based on young (less than 12 months) seedlings propagated in the nursery ('young' because ageing or maturation occurs early in life and negates the value of clonal BVs). Their breeding value can only be ascertained by several years of field testing of several ramets of a clone. There is no means of repropagating such clones from clonal tests while retaining their juvenile character. Selection of the best clones in these clonal tests can be used as a basis for mating them to create the next generation. They can be grafted into a CP orchard or propagated by grafting into an archive for mating, or the clonal test itself can be converted into a breeding archive or used for on 'ortet' pollination.

Within each subline, 24 parents will be used to produce the cloned families, and another 24 parents will be crossed to make the seedling families. Cloning is adopted to get maximum gains from traits of lower heritability, while the seedling families can be a source of selections for traits of higher heritability. There will be a total of 1920 cloned plants from 48 families per site in comparison with 960 seedlings from 48 families per site. Five sites will be planted with clonal tests and five sites for seedlings. It is intended that clones and seedlings will be tested together, with trials established every 2 years.

Deployment of the best material of the Elite might be through commercial varieties, but the Elite clones cannot be used for this directly because they cannot be rejuvenated. Propagating selected clones by grafting into a CP orchard is the simplest way of using them, with bulking up of resulting CP seed by cuttings. Converting a clonal test to a clonal orchard or planting the clones as a clonal test to be so converted is another possibility.

The cloned families resemble those of the cloned Elite in Shelbourne et al. (2007). The predicted gains from clones versus seedlings from this simulation were superior to seedlings, with heritabilities up to 0.5, and were only 1% (in 20%) less than seedlings at heritability 0.6.There will be a small number of crosses between sublines (about 10) as predeployment crosses of top-ranked parents with potential for commercial deployment of the offspring in CP seed orchards and clonal forestry.

**P.A. Jefferson 2016: The Breeding Management Plan (Radiata Pine Breeding Company Draft Report)**

This report is not a publication, and I was generously given a draft copy soon after the sudden death of Paul Jefferson, the author. This contains the latest description of NZ radiata pine breeding and is different from the strategy proposed by Dungey et al. (2009). This summary describes the development of radiata pine breeding from 2003 until 2016. I have edited and rearranged this material, but acknowledgment must go to P.A. Jefferson (2016).

NZ has long relied on progeny-tested clones in seed orchards for deployment of genetically improved seed of radiata pine. In spite of good gains from the development of control-pollinated orchards, there had been little progress in breeding population (BP) development and within-family selection. Selection of plus trees, collection of their OP seed and establishment of progeny trials and clonal seed orchards had gone some way to progress the previous breeding population. The status number had been substantially reduced by too much among-family selection.

A new breeding population was created by mating among the previous NZFRI clone series in the late 1980s and the seedlings planted at three sites in 1993 as the 'Big Bang'. Only one trial at Tarawera forest survived the attack of *Helicoverpa* caterpillars, and selections were made there by RPBC in 2003, and OP seed and scion material collected. Similarly, selections were made, and seed and scions were collected in the 1994/1995 planted wood density and *Dothistroma* Elite experiments in Kaingaroa ('2004' series), and in the '886' and '889' OP trials from unselected parents from North and South islands, also planted in Kaingaroa ('206' series). The new breeding programme is based on forwards selections from these as '2003', '2004' and '2006' series of families.

The parent breeding population is now 'stratified' into an Elite (top 1/3 BVs) and a Main BP (lower 2/3 BVs) on the BVs of the three series of progenies estimated at over a dozen sites of RPBC members. Some 360 forwards selections from the trials at Tarawera and Kaingaroa were made through standardised trial assessments, making the 96 '2003', 90 '2004' and 177 '2006' RPBC selections.

The current RPBC Elite and Main breeding populations are simply 'stratifications' of the OP progenies deriving from the most recent assessments of BVs of '2003', '2004' and '2006"series parents. The families with the best 120 (best 1/3) BVs are in the Elite, and those 240 families with the lower 2/3 BVs are in the Main BP. The whole 360 OP progeny BP is arbitrarily divided into two sublines. The '2003' s were originally selected for growth and form, the '2004's were half from a *Dothistroma*-resistant Elite and half from a wood density Elite, and the '2006' s were selected in an OP test of unselected material.

OP family tests of all 360 selections have been planted widely, and scions of their female parents have all been grafted at the Purchase road archive. The RPBC breeding population was derived from each series of selections ('2003', '2004' and '2006'). All '2003', '2004' and '2006' OP progeny have been tested widely in 11 trials throughout NZ.

All the '2003', '2004' and '2006' families have also been tested in Australia with 7 trials in NSW and 3 in Tasmania. The objectives are to rank them all for BVs for backwards selection for seed orchards, to investigate the impact of GxE on the sites and families and to provide BVs as the basis for crossing at Purchase road. None of the original OP seed of '2003', '2004' or '2006' selections was cloned.

To progress the BP, individuals from families may be cloned to provide better estimates of breeding values (BVs) and genotype x environment interaction (GxE) and to allow within-family clonal selection, with much higher heritabilities of clone means within families than within seedling families. Within-family selection on clone means will not reduce genetic variation in the BP as would selection among families. Clonal testing over a range of site types is currently being used in the crosses among the archived clones at Purchase road.

OP offspring from families planted over a decade ago provides estimates of breeding values across many sites in NZ and Australia, and since 2011 the RPBC has focused on selection traits DBH, wood density and stiffness.

Cloning seedlings of crosses in both Elite and Main BPs is being done in the breeding archive at Purchase road, Amberley. Time from initial selection to seed orchard deployment will be reduced from about 26 years to 16 years by clonal testing of seed from crosses among '2003' and '2004' series parents. The first cloned Elites of '2003' and '2004' selections were planted in NZ from 2013–2016 and in 2016 in Australia.

Cloning combined with within-family selection (no among-family selection) will allow selection of new BP members without loss of genetic variation and without inbreeding and excessive reduction in status number. If among-family selection in the BP was used, genetic variation would be much reduced for future selection.

As programmes progress, any new material will likely be inferior to the improved Elite and Main. This programme is effectively closed, and among-family selection would reduce genetic variation in the BP. However, a combination of within-family selection (which will not reduce genetic variation) and cloning will allow reliable forwards selection of BP clones as well as orchard clones and reduce time to deployment.

The parents of the 360 families, grafted at Purchase road, have been used to make crosses among Elite and Main population selections, either full-sib crosses or poly-crosses. Crossing of all selections to produce a new generation of the Elite and Main was done from 2007 to 2013, with more crosses in the Elite than the Main. The establishment of the first part of the new Elite over a range of sites was completed from 2013 to 2016.

(*Comment.* The amount of cloning of these families has not been noted in the draft BMP.)

Crossing involves the Elite and Main divisions of the '2003' and '2004' grafted parents. Some of these seedlings will be cloned for outplanting as the new BP.

(*Comment.* It appears that the CP seedling families are being planted widely, as for the OP families of the earlier generation, and that clonal tests will be planted on much fewer sites.)

The assessment of all 360 OP families on two sets of 11 NZ sites plus 7 Australian sites will generate a large data base for estimating genetic parameters and GxE variance. The BVs will enable reselection of the best clones, already archived at Purchase road for seed orchard planting. Controlled pollination and cloning will allow within-family selection of the future breeding population, as well as selection of the best clones for the CP orchards.

Crossing started in 2007 and continued till 2014, and CP tests will be established (2017–2022). At present, four lots of seed (of crosses among '2003' and '2004' selections) await cloning and planting in clonal tests. The last thing Paul J. wrote was, 'Each seedlot will be cloned, probably by 15 clones per family. These seedlots will be tested in clonal trials to maximise the value of….'

The Main BP parents will not be mated as intensively as those of the Elites. Also, the '2006' group of families were separated from the rest because of their generally lower BVs and placed in their own Elite and Main. New CP Elites of '2003' and '2004' were planted in NZ (13 sites) from 2013 to 2016, and in Australia in 2016 (4 sites).

Each Elite and each Main was divided into two sublines. Some of the key ideas in the paper could be identified as: 'the more advanced stages of tree improvement can be cloned to provide good estimates of BVs and GxE'. Heritabilities will be much higher at 0.2 and 0.3. In the RPBC programme, the most advanced generations and improved populations are currently using clonal testing over a range of site types. RPBC has now adopted clonal testing of Elite and Main populations. These gains can only be achieved by within-family selection. The progeny are well placed to maximise within-family selection. BPs are closed. Clonal testing of Main and Elite will maximise gains but new analytical methods will be needed.

Within-family selection in the BP will not reduce genetic variation, and also will barely increase inbreeding. Within-family selection is needed to maintain genetic diversity in the BP, and cloning of families is ideally suited for doing this. Within-family clonal selection can be unrestricted and does not reduce genetic variation. Clonal within-family selection will increase gains in both BP and PP.

## References


I.J. Thulin 1957: Application of tree breeding to New Zealand forestry. NZ Forest Service Tech. Pap. 22 (1957)

C.J.A. Shelbourne 1969: Tree breeding methods. Technical Paper 55, Forest Research Institute, New Zealand Forest Service, Wellington

41. C.J.A. Shelbourne and F.R.M. Cockrem 1969: Progeny and clonal test designs for New Zealand's tree breeding programme

R. 772. C.J.A. Shelbourne 1973: Problems and prospects in the improvement of forest tree species. Proc. 2Nd General Congress, SABRAO, New Delhi 1973. reprinted from Indian J. Genet., 34A 1974

R. 531. R.D. Burdon and C.J.A. Shelbourne 1971: Breeding populations for recurrent selection: conflicts and possible solutions. NZJFS 1(2):1174–193

R. 548 C.J.A. Shelbourne 1971: Planning breeding programs for tropical conifers grown as exotics. IUFRO (section 22) Gainesville, Fla. 1971. Symposium on "Selection and breeding to improve some tropical conifers"

R. 683 C.J.A. Shelbourne 1972: Genotype-environment interaction: its study and its implications in forest tree improvement. IUFRO Genetics-Sabrao Joint Symposia.Tokyo, October 1972

R. 1021. R.D. Burdon, C.J.A. Shelbourne and M.D. Wilcox 1977: Advanced selection strategies. Third World Consultation on Forest Tree Breeding, Canberra and Rotorua

R. 1089. R.D. Burdon 1977: Genetic correlation as a concept for studying genotype-environment interaction in forest tree breeding. Silvae Genetica 26(5–6):145–228

R. 1296. R.D. Burdon 1981: Generalisation of multi-trait selection indices using information from several sites. NZJFS 9(2): 145–152

R. 1518. R.D. Burdon 1982a: Selection indices using information from multiple sources for the single trait case. Silvae Genetica 31(2–3): 81–85

R. 1599. R.D. Burdon and G. Namkoong 1983: Short note: Multiple populations and sublines. Silvae Genetica 32(5–6):221–222

R. 1640. R.D. Burdon 1982b: Breeding for productivity- jackpot or will-o-the-wisp? Proc. North American Forest Biology Workshop, University of Kentucky, Lexington, Ke., July 1982

R. 1736. G.B. Sweet, and R.D. Burdon 1983: The radiata pine monoculture: an examination of the ideologies. NZ J. For. 28(3): 325–26

D.V. Shaw and J.W. Hood 1985: Maximising gain per effort by using clonal replication in genetic tests. TAG 71:392–399

C.J.A. Shelbourne, R.D. Burdon, S.D. Carson, A. Firth and T.G. Vincent 1986: Development plan for radiata pine breeding. Forest Research Institute, Rotorua, New Zealand

R.D. Burdon 1986: Breeding long-lived perennials-frustration, temptations, opportunities. Plant Breeding Symposium DSIR 1986 Agron. Soc. NZ special pub. no. 5

R.D. Burdon 1989: Early selection in tree breeding: principles for applying index selection and inferring input parameters. Can. J. For. Res. 19: 499–504

C.J.A. Shelbourne 1992: Genetic gains from different kinds of breeding populations and seed or plant production populations. S. Afr. For. J. 160:49–6

R. 2453. M.J. Carson, T.G. Vincent and A. Firth 1992: Control-pollinated and meadow seed orchards of radiata pine. Mass Production Technology for Genetically Improved Fast growing Forest Tree Species. Bordeaux 14–18 Sept. 1992. AFOCEL Paris 1992

R. 2503 R.D. Burdon 1992: Testing and selection: strategies and tactics for the future. pp. 249–60 Proc. IUFRO Conf. Oct. 9-18 1992, Catagena and Cali, Colombia. Sponsored by CAMCORE

T.L. White, G.R. Hodge and G.L. Powell 1999: An advanced genetic improvement plan for slash pine in the southeastern United States. Silvae Genetica 42 (6): 359–370

R. 2992 R.D. Burdon 1995: New directions in tree breeding: some questions of what we should seek, and how to manage the genetic resource. CTIA/WFGA meeeting, Victoria, BC August 1995

R. 2659. L.D. Gea, D. Lindgren, C.J.A. Shelbourne and T. Mullin 1997: Complementing inbreeding coefficient information with status number: implications for structuring breeding populations. NZJFS 27(3): 255–271

C.J.A. Shelbourne, L.A. Apiolaza, K.J.S. Jayawickrama and C.T. Sorensson 1997: Developing breeding objectives for radiata pine in New Zealand. Proc. IUFRO '97 Genetics of Radiata Pine, Rotorua, NZ, Dec. 1–4 1997

O. Rosvall, D. Lindgren and T.J. Mullin 1998: Sustainability, robustness and efficiency of a multi-generation breeding strategy based on within-family clonal selection. Silvae Genetica 47 (5/6): 307–320

R. 2745. R.D. Burdon 2001: Genetic aspects of risk: species diversification, genetic. management and genetic engineering. NZ J. Forestry, Feb. 2001:20–25

S. Kumar 2006: Correlation between clonal means and seedling progeny means, and its implications for radiata pine breeding strategy. Canadian Journal of Forest Research 36:1968–1975

R. 2842. R.D. Burdon and S. Kumar 2004: Forwards versus backwards selection: trade-offs between expected genetic gain and risk avoidance. NZJFS 34(1): 3–21

R. 2847. R.D. Burdon, R.P. Kibblewhite, J.F. Walker, R. Evans and D.J. Cown 2004: Juvenile versus mature wood: a new concept, orthogonal to corewood versus outerwood, with special reference to *Pinus radiata* and *P. taeda*. Forest Science 50(4): 399–415

H.S. Dungey, J.T. Brawner, F. Burger, M.J. Carson, M. Henson, P.A. Jefferson and A.C. Matheson 2005: A new breeding strategy for the Radiata Pine Breeding Consortium: summary and outcomes of the Noosa workshop. RPBC Report?

39460 C.J.A. Shelbourne, R.D. Burdon, H.S. Dungey, L.D. Gea and S. Kumar 2006: Revisiting the elite. RPBC/Ensis report? Brain-storming workshop 28$^{th}$ February 2006

C.J.A. Shelbourne, S. Kumar, R.D. Burdon, L.D. Gea and H.S. Dungey 2007: Deterministic simulation of gains for seedling and cloned main and elite breeding populations of *Pinus radiata* and implications for strategy. Silvae Genetica 56: 253–300

38627 R.D. Burdon 2005: The RPBC breeding strategy: profit chasing and risk management

39461 H.S. Dungey 2006a: Revisiting the elite-crossing and selection. RPBC/Ensis report? Brainstorming workshop 28$^{th}$ February 2006

37297 T.G. Vincent 2006: Review of genetic resources of introduced tree species other than PR in NZ

P. Cotterill 1986: Breeding strategy: Don't underestimate simplicity. IUFRO Working Parties on Breeding Theory, Progeny Testing and Seed Orchards. Williamsburg, Va. USA. October 13 1986.

H.S. Dungey, J.T. Brawner, F. Burger, M. Henson, P.A. Jefferson and A.C. Matheson 2009: A new breeding strategy for *Pinus radiata* in New Zealand and New South Wales. Silvae Genetica 58, 1–2 (2009)

P.A. Jefferson 2016: The Breeding Management Plan (Radiata Pine Breeding Company draft report)

K.J.S. Jayawickrama and C.B. Low 1999: Pinus radiata selections from different regions in NZ 1250 differ in branch habit, form and growth rate. NZ J. For. Sci. 29(1) 3–21.

Namkoong, G. 1966: Family indices for seed orchard selection. Joint Proceedings of Second Genetics Workshop and Seventh Lake States Forest Tree Improvement Conference, United States Forest Service Research Paper NC-6: 7–12.

K.J.S. Jayawickrama and M.J. Carson 2000: A breeding strategy for the New ZealanRadiata Pine Breeding Cooperative. Silvae Genetica 49, 2.

White, D. A., Raymond, C. A., Kile, G. A., & Hall, M. F. (1999). Are there genetic differences in susceptibility of Eucalyptus nitens and E. regnans stems to defect and decay? Australian Forestry, 62(4), 368–374.

R. 2367. M.J. Carson, R.D. Burdon, S.D. Carson, A. Firth, C.J.A. Shelbourne and T.G. Vincent 1990: Realising genetic gains in production forests. IUFRO Working Parties on Douglas fir, Lodgepole pine Sitka spruce and Abies spp. Breeding and genetic resources. Session: Genetic Gains in production forests.

# Chapter 15
# Conclusions

## 15.1 Chapters 1–14

The provenance testing results of the different species from Europe, Japan, Eastern and Western America and Australia showed that Douglas-fir was the only species of conifer to prove acceptable. *Eucalyptus fastigata, E. nitens* in Southland and *E. regnans* are also in breeding programmes. The historical development of tree breeding of radiata pine was traced from the intensive plus-tree selection of the 1950s to the more recent programme of the 1990s with two superlines and four small Elites. The most recent move is to abandon the special purpose Elite populations and use Kinghorn algorithms to select the BP and PP.

This book is primarily a chronological 'compendium' of summaries and abstracts of most of the published material produced by NZ tree breeders, with my commentary adjacent. A large number of internal reports, as listed, were not made available as PDFs by Scion for commercial competitive reasons.

Species and provenance research is featured from an era when a wide range of species were still under consideration. The effects of *Dothistroma pini* and the opossum were quite catastrophic for some species, yet radiata pine withstood them to a large extent, with human help. New diseases Phytopthera, pitch canker and Western gall rust are future threats.

Provenance trials of *P. elliottii* (slash) and *P. taeda* (loblolly) were among the first and were quite well adapted to phosphate-deficient gumland clay sites. Their low wood density and, for *P. taeda*, bark inclusions beneath branches and the severe damage caused by possums were eventually damning.

Corsican pine, *P. nigra*, a very widespread and important species, showed a clear provenance story, but *Dothistroma* hit the desired Corsican provenance which couldn't be grown anywhere except in dry, cold South Island sites. The depression-planted stands of *P. ponderosa* var. *scopulorum* were horrible examples of the wrong provenance. By 1960 *P. ponderosa* provenance research was well advanced, showing the good performance of the provenance of Grants Pass in Oregon, but



C. J. A. Shelbourne, M. Carson, *Tree Breeding and Genetics in New Zealand*,
https://doi.org/10.1007/978-3-030-18460-5_15

*Dothistroma* needle-cast put paid to any future for this species. *P. pinaster*, maritime pine from Portugal and SW France, grew better than the rest of wide-ranging provenances, the Corsican being the straightest but slower than the not-so-well-formed Portuguese and French Landes provenances. This was once a useful species for phosphate-deficient sites but was displaced by good land preparation and fertiliser which favoured radiata.

West Coast American species were generally better adapted than those from other regions. *P. contorta, Pseudotsuga menziesii, Picea sitchensis, Tsuga heterophylla, Thuja plicata, P. muricata* and *P. attenuata* were all tested, but, except for Douglas-fir, most didn't make the grade, mainly on growth rate. In the end, disease and species inferiority to radiata pine in growth, form and adaptability narrowed the list to Douglas-fir, and certain eucalypts.

The breeding strategy for Douglas-fir demonstrated the critical importance of provenance. Based in 1968 on selection in the extensive Kaingaroa stands of probable Washington origin, it became clear that the southern coastal provenances performed so much better. Provenance mistakes, easy to make, were thus built into the Douglas-fir breeding strategy, underlining the need for good provenance work before starting selection. The Douglas-fir breeding programme was put on hold for over 20 years, but in 1988 it was resurrected by making 180 selections from the best provenances in the 1959 trials.

Eucalypt provenance and breeding work started in 1975, and the species involved were *E. regnans*, *E. fastigata*, *E. nitens*, *E. saligna* and *E. delegatensis.* As the trials were designed to run for a limited period, early assessment of frost damage and malformation made early choice of progenies and provenances necessary. The programme was often carried out in too short a period and in too much of a hurry. An important rule in designing a breeding programme is to design for final assessment up to half a rotation ahead. Eucalypts are notorious in changing their performance with age. Some interest has arisen in *E. bosistoana* for durable vineyard posts and in stringybarks *E. muelleriana* and *E. globoidea* for sawtimber. A lot of in-depth research has been done using 29-tree samples on kraft pulping of *E. regnans, E. fastigata* and *E. nitens* and on 9 trees of these species for mechanical pulping, showing their pulps were unfortunately low on bulk.

The radiata pine breeding strategy has been a principal focus of this book. An interesting aspect is the development of this strategy over such a long period, some 65 years. Ib Thulin started the breeding programme in 1952, before the industrial breeding programmes were begun at North Carolina State, University of Florida, College Station, Texas, and before any Douglas-fir breeding got started in BC or Washington.

For the first 15 years or more, there was little change in the radiata breeding strategy, simply very intensive plus-tree selection in the virtually unmanaged stands of radiata pine in Kaingaroa forest, with clonal seed orchards. There was little attention given to any systematic progeny testing, though an OP test of 29 families and another of some CP families were very useful to breeders 12 years later.

The next step was a drastic change. About 600 plus trees were selected much less intensively from young stands, 12–17 years old, and OP tested on 3 sites.

Seed orchards were later planted with progeny-tested clones. The sets in reps design of Schutz and Cockerham was firmly adopted. A group of about 100 '875' trees were later forwards selected from the '268' OP families, and these were crossed by climbing the parent 875s. They were not suited to AS as they were from only the best 90 of the 588 '268' families, though this was not realised at the time.

In 1986 the Development Plan stimulated thinking about how to breed the species. The initiation of control-pollinated seed orchards was an important new factor. Some breeders finally realised that within-family selection for the next BP was needed for AS, rather than among-family selection, which had reduced genetic variation in the BP and reduced its status number. Cloning within each AS family was needed to increase heritability from that of phenotypic selection, but rooting of seedlings had yet to be developed. Advanced generation selection had relied in the past on backwards selection, so that the best clones of the big OP population were still crossed for planting for AS, really a very bad strategy which caused major reduction of genetic variation in the breeding population.

Later, deterministic simulation showed that cloning in a balanced (within-family) selected population gave much better gains than seedlings. It was also later shown that clone means of juvenile cuttings correlated well with seedling breeding values.

The RPBC is now running the breeding programme and has tested 360 OP families, forwards selected from the Big Bang, *Dothistroma* and high wood density elites, and another OP test, to test the genetic base over 11 sites in NZ and Australia. They have crossed selections (as grafts) from these series and are cloning some seed from each cross. They have apparently embraced the idea of conserving genetic variability in their breeding population (which is effectively closed) with use of within-family selection combined with cloning. Their breeding strategy is where we should have been 30 years ago.

At a recent (2018) review meeting, it was agreed to abandon the 24 parent Elite populations, and the division into two superlines was also abandoned to make a single breeding population. This would be on the basis of using Kinghorn algorithms to control relatedness and to effect selection for BP and PP.

And this really wraps up the RP breeding strategy story as detailed in this book. The RPBC people are doing it about right. It is just such a pity that Paul Jefferson is not here to see it.

# Chapter 16
# The Future of Forest Tree Improvement in New Zealand

## 16.1 Summary

Clonal testing for within-family 'forwards' selection has begun in the RPBC breeding programme and should also provide a powerful means of increasing gains in the breeding population. Within-family selection does not reduce breeding population genetic diversity, like selection among families.

NZ wood and wood products need innovation to move radiata pine wood from commodity to higher value uses. Economic breeding objectives are needed in which the aggregate genetic value improvement is described in dollar terms. However, as James pointed out, in the long term, a single breeding objective can be the most efficient.

Quicker selection is a key target since it can enable higher genetic gains per year in both the breeding and deployment operations (improving gain per year is a key objective for this program). OP breeding methods can shorten the breeding cycle by saving time in the progeny testing stage while achieving almost similar levels of genetic gain to those from a control-pollinated (CP)-based program. OP seed collected from plus trees and OP forwards selections can be used for progeny testing, obviating grafted archives for subsequent CP development.

At a recent review, the previously planned Elite, Main and Superline structures have been combined into a single breeding population, utilising mate selection by algorithms developed by Kinghorn and others. Clonal testing of candidate for forwards selections using rooted cuttings allows within-family selection without loss of genetic diversity in the BP. It also provides benefits of higher heritability, leading to higher genetic gains/year in PP and BP. Improved confidence in EBVs (estimated breeding values) for the cloned forwards selections has enabled seed providers to begin routinely grafting cloned forwards selections into clonal seed orchards.

Efficiency gains come from improved methods of breeding value estimation and improvements in progeny trial designs such as 'multiple environment trial' (MET)



C. J. A. Shelbourne, M. Carson, *Tree Breeding and Genetics in New Zealand*,
https://doi.org/10.1007/978-3-030-18460-5_16

analysis methods for radiata pine, using the BLUP reduced animal model, combined with factor analysis.

Recent work has demonstrated wood property screening of tree stem and core samples from trees as young as age 2, using new laboratory techniques. Traits screened with these tools include wood basic density, spiral grain, resin content and compression wood. Tree improvement programmes will face new challenges from pests and diseases. New Zealand's exotic forest plantations are losing their benefits of remoteness as new exotic pests and diseases continually arrive. The implementation of genomic selection (GS) is becoming a key issue for tree breeding programmes worldwide, as technical development challenges are being met.

Implementation of GS will mainly be for earlier selection than can be achieved using current screening methods. Such early selection is particularly valuable for economically important traits like cell-wall galactans and slow-to-develop heartwood. Earlier selection for low heritability traits like growth rate will also be valuable in shortening the progeny testing phase. The transfer of GS-based genetic gains into plantations will be most efficient and effective using somatic clones (due to their higher trait heritabilities), less so for CP seedlings and cuttings, and substantially less for OP seedlings. Cloned forwards selections cannot be multiplied for deployment as they are irreversibly matured during progeny testing.

Somatic cloning (through seed-based embryogenesis and cryogenic storage) is one method of cloning the breeding population, which could lead to a 'step-change' in delivery of genetic gains, in spite of only 10–15% of seeds and few families yielding somatic clones. Somatic clones, after field testing, can be immediately mass-propagated from cryogenic storage and deployed in plantations which would allow both higher genetic gain and increased gains/year from the breeding programme. Direct deployment of somatic clones is the best means of using genomic selection, as well as CRISPR-type gene editing and gene transfer methods.

## 16.2 Introduction

As for all major tree improvement programmes, the future of NZ tree improvement will be based on the population structures and levels of genetic gain achieved from previous generations of tree breeding (Shelbourne et al. 1986; Jayawickrama and Carson 2000; Dungey et al. 2009). The plantation estate is dominated by radiata pine, which will continue to provide the model for future tree improvement. The breeding programme for radiata pine is managed by the Radiata Pine Breeding Company (RPBC) which supplies treestocks from seed and somatic clones to the NZ pine estate of over 1.5 million hectares, as well as some 200–300,000 hectares of the Australian plantation estate.

By comparison, plantation areas and the related tree improvement efforts in NZ for other species are small. NZ Douglas-fir plantations have peaked at around 100,000 hectares, all eucalypts total approximately 23,000 hectares, with cypresses and redwoods next at approx. 10,000 and 8000 hectares, respectively. Planting of

New Zealand's indigenous species has historically been negligible, although the recent emergence of NZ manuka (*Leptospermum scoparium*) as a producer of honey for both consumption and as a component of various health products has led to increased plantings. Similarly, a combination of renewed Maori interest and the current government's 'Billion Tree' policy may lead to a surge in planting of indigenous species.

The available genetic diversity in the radiata pine breeding programme is greatest in the 'Gene Resource Population', which is comprised of the landraces established in NZ since the 1860s, and stands of the five native provenances from California and Mexico. The 'breeding population' has been formed from plus-tree selections, originally from the landraces, plus limited infusions of native provenances (principally from Guadalupe Island).

The 'production population' comprises the progeny-tested parent clones established in open-pollinated and control-pollinated seed orchards and contains less genetic diversity but represents substantially greater gains from selection. Finally, the 'deployment population' represents the point of delivery for all genetic improvement efforts and is comprised of the highly selected and improved seed orchard seedlots plus individual improved somatic clones that are established in forest plantations (as yet, only 10–15% of seeds and few families yield somatic clones). In this population, genetic gain is maximised at the expense of reduced genetic diversity and potentially increased levels of inbreeding, and the trade-offs among these factors become important.

## 16.3 Population Structure and Development

The population structures described above are likely to be maintained in future tree improvement, as will the continuing use of classical breeding methods based on quantitative genetic theory, which will be increasingly informed by genomics tools. Until very recently, the RPBC breeding population has been further structured into a small Elite population and a larger Main population, with both populations divided into two superlines, in order to delay inbreeding by facilitating outcrossing in the deployment population (ex Dungey et al. 2009). However, a major recent review has resulted in the decision to combine the Elite, Main and Superline structures into a single breeding population and to utilise mate selection for achieving assortative mating, using efficient algorithms developed by Kinghorn and others (1999). These changes are expected to achieve the necessary optimisation of breeding gains and control of inbreeding versus any associated loss of genetic diversity from the breeding population. Clonal testing for within-family 'forwards' selection, which was not feasible in the past, has commenced in the RPBC breeding programme and should also provide a powerful means of increasing gains in the breeding population.

Tree breeders traditionally have defended the maintenance of large breeding populations, and, if costs are not an issue and everything else works as it should, few would argue against this.

Arguments for a large breeding population have included:

- Needed to ensure access to most of the species' available genetic diversity/variance
- Enables higher selection intensities to be used in each breeding and selection cycle, therefore potentially achieving higher gain
- May protect against loss of rare alleles (e.g. for disease resistance)
- Delays the onset of relatedness and associated risks from inbreeding

However, recent studies based in part on population simulations support the use of much smaller breeding populations in line with practice in numerous other plant crops. For example, simulation work by Wu et al. (2016) indicates that maintenance of a per generation population size of around 200 individuals can be sufficient to balance the genetic gain versus diversity equation.

A rolling front approach, defined as '… a strategy in which all the main breeding activities are undertaken on a yearly basis' (Borralho and Dutkowski 1998), has become the standard approach used in advancing the Australian STBA tree improvement programme and has been recently adopted for the NZ programme. Each year parents are crossed, new progeny field trials are established, some existing trials are measured, new data are incorporated into the genetic evaluation system, and new parents are selected on the basis of the updated genetic values.

## 16.4 The Breeding Objective

The challenge of maintaining international markets for NZ wood and wood products will continue to require innovation as the forestry sector attempts to increase value-add by moving radiata pine wood from commodity to higher value uses. For the NZ radiata pine breeding programme, this necessitates use of a well-defined economic breeding objective, in which the aggregate genetic value of desired improvement is described in dollar terms (Alzamora 2010). The definition of genetic merit can vary along the production and processing chain. For example, Ivkovic et al. (2010) showed that the relative values of mean annual increment, sweep, branch size and modulus of elasticity vary between growers, sawmillers and integrated enterprises. However, James (1987) pointed out that, in the long term, a single breeding objective can be most efficient, and Apiolaza and Garrick (2001) have shown this to be appropriate for NZ radiata pine. A current NZ breeding objective will be likely to place most emphasis on improved growth rate (and/or associated reductions in rotation length), with secondary emphases on wood quality (strength and stiffness), tree form (stem straightness and reduced leader malformation) and resistance to Dothistroma.

Today's economic values may not adequately reflect those prevailing when wood harvest will occur several decades into the future. To combat this risk, Evison and Apiolaza (2014) have outlined a method (termed 'robust selection') for dealing with market uncertainty through selection for individuals expressing less variability in their expected economic value, as estimated from a multi-trait selection index.

## 16.5 Improved Breeding Efficiency

Other proposed changes to the breeding strategy are aimed at achieving improvements in overall efficiency, plus increasing gains per year through earlier selection (including genomic selection) and more sophisticated methods of seed and plant delivery.

Earlier (and often, indirect) selection is a key target for all tree improvement programmes, since it can enable achievement of higher genetic gains per year in both the breeding and deployment operations (King and Burdon 1991; Kumar and Lee 2002; Li and Wu 2005). For the NZ radiata pine programme, the choice of the minimum selection age in progeny trials has been largely set as the earliest age at which the breast height diameter (DBH) can be considered to be a reliable estimator of rotation-end tree volume. In recent years, the practice has been to select at age 8 from trial establishment, by which time both the expressions of genetic variance and heritability for DBH are close to their maximum and the trial trees have not yet been strongly affected by inter-tree competition effects due to canopy closure. By age 8 also, the key form traits of stem straightness and branch cluster frequency (and associated internode length) can be easily assessed (Carson and Inglis 1988) and also the key wood property traits of wood density and corewood stiffness (Apiolaza et al. 2008; Apiolaza 2009; M. Lausberg, personal communication) and resistance to the Dothistroma needle blight (Carson 1989).

More efficient and often earlier selection will be achieved in a number of ways, including through use of:

- Open-pollinated (OP) breeding methods (Shelbourne et al. 2007)
- Clonal testing, combined with forwards selection
- Improved breeding value estimation and progeny trial designs (Cullis et al. 2014)
- New wood property screening methods (Apiolaza et al. 2008)
- Remote sensing applications
- Databases and decision support tools
- Screening for disease resistance
- Applications of genomics methods

## 16.6 Open-Pollinated (OP) Breeding Methods

OP breeding methods can shorten the breeding cycle by saving time in the progeny testing stage while achieving almost similar levels of genetic gain to those for a control-pollinated (CP)-based programme. OP seed collected from plus-tree and OP forwards selections (assumed half-sib in both cases) can be used for progeny testing, thereby saving the time taken to establish and achieve flowering in grafted archives for subsequent development of full-sib crosses using controlled pollination. While the associated loss of pedigree information on the male line has been of concern in the past, genotyping tools are now available to identify relatedness among OP trees in most situations. For the present, however, the NZ radiata pine breeding programme is continuing to rely on fully pedigreed CP crossing for breeding population development. The additional time needed for pursuing a CP breeding strategy can be reduced if selection and CP crossing take place in the year of final progeny trial assessment; achieving this is now a practical goal for the NZ breeding programme.

## 16.7 Clonal Testing with Forwards Selection Within Families

Clonal testing of candidate for forwards selections using rooted cuttings is now an accepted practice in the NZ RPBC programme, providing benefits of higher heritability, increased genetic diversity in the BP and associated greater confidence in estimated breeding values (EBVs), leading to higher genetic gains/year. Cloning of radiata pine can be achieved through any of the following:

Grafting of scions from both juvenile and mature trees, for establishment of clonal seed orchards and genetic archives

Rooting cuttings from juvenile trees, usually combined with hedging to delay the effects of physiological aging and used routinely for multiplication of CP cross treestocks for deployment.

Tissue culture, usually shoot cultures of axillary bud tissue from juvenile plants, used currently for an intermediate bulking of somatic cell lines prior to their establishment in cuttings stoolbeds.

Somatic clonal lines developed from immature seed and cryostored in liquid nitrogen in order to delay the negative effects of physiological aging (as yet, only 10–15% of seeds and few families yield somatic clones). Somatic clones being deployed in NZ represent approximately 8% of the annual planting programme.

Cloned forwards selections from progeny trials are sexually mature by around age 4–5 from planting, and this currently sets a lower limit on the time required for completion of a breeding cycle. When applied appropriately, top grafting of material from juvenile selections onto mature plants offers opportunities to reduce this age by 1–2 years (Li and Dungey 2018). Either way, the more accurate EBVs estimated for cloned selections, replicated as they are both within and across trial sites, provide added confidence that can be translated into a reduction of 2 or more years of the current 8 years required for progeny testing.

The same improved confidence in EBVs for the cloned forwards selections has enabled seed providers to begin routinely grafting them into clonal seed orchards earlier to produce commercial seed for treestock deployment. Both OP and CP seed are produced in these orchards, with the CP component usually being deployed in the form of rooted cuttings from hedged stoolbeds of CP seedlings. Somatic clones are similarly deployed as rooted cuttings, following a tissue culture bulking up phase, and continued supply of these is made possible through replacement of aging stoolbeds with new juvenile stocks extracted from cryostorage.

## 16.8 Improved Breeding Value Estimation and Trial Designs

Efficiency gains are coming from improved methods of breeding value estimation and improvements in progeny trial designs. University of Wollongong researchers have developed 'multiple environment trial' (MET) analysis methods for radiata pine, using the BLUP reduced animal model, combined with factor analysis (Cullis et al. 2014). This work has provided new selection tools for separately characterising and selecting candidates for both their overall performance and stability of performance across sites, thereby offering opportunities to better manage G X E and to achieve improved genotype/site matching (or 'G + E') (Smith and Cullis In Press). New 'sparse phenotyping' progeny trial designs promise to further reduce costs of investment in trials while also improving trial coverage (B. Cullis, personal communication).

## 16.9 New Wood Property Screening Methods

Recent work at the University of Canterbury, School of Forestry, has demonstrated the potential for wood property screening of tree stem and core samples from trees as young as age 2, using new laboratory techniques (Apiolaza et al. 2008). Traits screened with these tools include wood basic density, spiral grain, resin content and compression wood. These methods should enable earlier and more intensive screening of selection candidates.

Trial assessment of corewood stiffness and wood density with new tools like the Resistograph may prove cost-effective and can be used in combination with the existing 'time-of-flight' Treetap and Hitman tools (M. Lausberg and G. Downes, personal communication).

## 16.10 Remote Sensing Applications

Remote sensing applications of tools like LIDAR ('Light Detection And Ranging'), and UAVs ('Unmanned Aerial Vehicles') are enabling more cost-efficient phenotyping of genotypes to be achieved in both trials and forest stands. LIDAR can already provide cost-effective data on trial plant survival and trial heights, as well as disease traits and nutritional status. Similar data collected on forest stands will enable tracking of the plantation performance of improved family and clonal treestocks, providing early feedback for guiding decision-making in both the NZ RPBC breeding and deployment populations. In addition, these tools will provide early screening opportunities against new pest and disease incursions.

## 16.11 Databases and Decision Support Tools

Tree breeding programmes have increased in their complexity with increasing use of sophisticated trial designs and both multisite and multi-trait analysis methods and the extended pedigrees resulting from progressive breeding cycles. Various computer-based tools are now evolving to enable the NZ breeding programme to be managed more efficiently, including the Gemview (www.gemnetics.com) and Katmandoo relational databases. In addition, forest growers are being provided with decision support tools for optimising deployment of improved genotypes, with current examples including the RPBC's GenSelector software and a Univ. Wollongong application for use with mobile devices.

## 16.12 Screening for Disease Resistance

While the implementation of these new technologies will bring welcome increases in forest productivity, tree improvement programmes will also face new challenges from pests and diseases. New Zealand's exotic forest plantations are progressively losing the benefits of remoteness that they previously enjoyed, as new exotic pests and diseases continually arrive and threaten plantation health and productivity.

Such incursions may further increase as a result of climate change. While conventional field screening has been very effective for selection of Dothistroma-resistant trees in NZ (Carson and Carson 1989), research is continuing into development of both field and indirect screening methods for other foliar diseases, for example, the recently identified red needle cast (RNC) disease (Dungey et al. 2014). Pre-emptive resistance screening is often prescribed for protection against diseases that may not have yet arrived, but have proven damaging in other countries. For example, tree breeders in the USA, Chile, Australia and NZ combined to complete a pre-emptive screening project that has provided preliminary resistance scores to the pine pitch canker disease in radiata pine, which is endemic in the US Pacific NW (Matheson et al. 2006).

## 16.13 Applications of Genomics Methods to Enhance Breeding

The implementation of genomic selection (GS) is becoming a key issue for tree breeding programmes worldwide as technical development challenges are being met (Resende et al. 2012; Isik et al. 2016; Grattapaglia 2017). NZ R&D providers are close to completing a major project that will deliver a single nucleotide polymorphism (SNP) panel, capable of being used for GS in both the radiata pine breeding and deployment programmes. Current costs are estimated at approx. NZ$37 and potentially as low as NZ$25 per genotype for the SNP-based GS test, depending largely on scale (J. McEwan, personal communication).

Although there are challenges in terms of both biological and cost constraints to GS implementation, reports of successful progress with GS in relatively small breeding populations in other crops have been encouraging, and statistical methods for combining GS and field measurement data are continually being improved. Implementation of GS will principally be for earlier selection than can be achieved using current field and greenhouse screening methods. Such early selection is particularly valuable for economically important traits that either are very expensive to measure using conventional methods (e.g. cell-wall galactans) or take a long time to develop in field trials (e.g. heartwood development). However, even earlier selection for more conventional, low heritability traits like growth rate will also be valuable in shortening the progeny testing phase.

In a radiata pine breeding programme, a new selection cycle/breeding generation might be accomplished using GS within a 7–10-year timeframe, instead of the current 12–15-year breeding cycle using conventional field testing. GS may also be used for rapid improvement of specific key traits and/or for a potential emergency response to a new pest or disease threat, enabling repeated generation cycles to be implemented for single traits, leading to accelerated breeding gains/year for that trait.

## 16.14 Applications of GS to Enhance Deployment

The implementation of GS for improving commercial treestocks which are directly deployed by growers of radiata pine represents a much greater challenge than that for breeding. The transfer of GS-based genetic gains into plantations will be most efficient and effective using somatic clones (due to their higher trait heritabilities), less so for CP seedlings and cuttings, and substantially less for OP seedlings.

The cloned forwards selections selected in the breeding programme cannot currently be multiplied for deployment since they will have undergone irreversible maturation during the progeny testing and reproductive development phases. New CP cross seed derived from breeding population parents can be screened using GS, but then there will be a requirement for the selected set of individual genotypes to be massively multiplied (using some combination of tissue culture bulking and stoolbed cutting production) in order to create a sufficient number of plants for establishment of meaningful areas of commercial plantations. This will inevitably involve a trade-off between achieving worthwhile selection intensity and incurring costs for a large number of SNP reactions, and these costs will be further increased by the requirement to turn cuttings stoolbeds over every 4 years, in order to control maturation.

NZ clonal providers have a resource of several thousand improved clones that have been captured into somatic embryogenesis and cryostorage. These represent an efficient and effective opportunity for applying GS to realise higher genetic gains for both breeding and commercial deployment. Plant production costs will, of course, continue to be higher for somatic clones and CP treestocks versus OP seedlings, with today's costs being in the order of $700/thousand for somatic clones, $450/thousand for CP cuttings, $350/thousand for CP seedlings and around $250/thousand for OP seedlings. However, the gains from clonal deployment plus genomic selection will outweigh the additional costs of production.

There are additional potential benefits from the use of genomic methods beyond implementing GS selection, including:

Breeding values can be estimated more accurately, through reductions in identification errors, plus better estimates of relatedness from the additional genomic information.

Improved accuracy in genotyping should assist in improved quality control in breeding operations.

Genomic data can play an essential role in augmenting any future applications of genetic editing and manipulation techniques.

SNP associations with key traits can be used to identify higher gain candidate selections for breeding crosses, through improved detection of both favourable gene clusters and potential inbreeding.

## 16.15 Realised Gains and Tree Crop Modelling

Proof of profit is an essential prerequisite to market demand for improved treestocks, and most forest growers in NZ believe that the continued quantification of genetic gain is essential. Forest growers use genetic gain information in new crop investment decisions as to what treestocks to plant, in yield tables and in determining woodflow and harvest levels.

Tree breeders have sometimes relied on predicted genetic gains from breeding values developed from progeny trial data to be realistic estimates of realised genetic gain. For example, the STBA uses this method to estimate annual gains in their programme (Powell et al. 2004). However, the main limitation of this approach is that it does not take adequate account of the relatively major (compared to genetics) effects of site and silviculture (Carson et al. 1999a).

The much greater impacts of site and silviculture (especially on tree growth) mean that these can have a large influence on the expression of genetic gain, which is not reflected in breeding value estimates. Instead, the use of well-designed and forward-looking long-term large-block trials combined with tree crop modelling is considered to be a more reliable approach (Carson et al. 1999b).

Tree crop models developed in NZ for growth rate, wood density and corewood stiffness are based on thousands of empirical measurements over a comprehensive range of forest sites. Early-age breeding values 'grown on' in these models are far more likely to yield reliable estimates of rotation-end values that can be used in genetic gain estimation, since they will take due account of site and silviculture effects. In addition, age/age genetic correlation estimates derived from progeny trial data are used in breeding value estimation and selection indices for key traits like growth rate. While conventional single-tree-plot progeny trials are very efficient at ranking genotypes for selection, they are as poor for providing reliable age/age correlation estimates as they are for estimating realised rotation-end gains. Instead, breeders may better accept the age 8 breeding values from progeny trials as providing the best- available information on genotype ranking for selection and use the growth (or other target trait) models to provide the necessary selection trait correlation with the requisite (and often quite different) rotation-end target trait (Carson et al. 1999b). The theoretical counter to this would be to challenge the implicit assumption that this approach makes, that is, that there is no crossover of genotype rankings between age 8 and age 25–30. However, we would argue that this is a safer option than the use of very unreliable age/age correlation estimates derived from progeny trial data.

Long-term 'genetic gain trials' comparing seedlots with various levels of improvement were first established in NZ in 1978 (Shelbourne et al. 1986). These trials were subsequently enhanced with the establishment of a comprehensive series of 'silviculture/breed trials' between 1987 and 1992 (Carson et al. 1999a), aimed at providing data on genetic performance of a mensurational standard over a range of silvicultural (largely tree spacing and thinning) options and across all NZ growing regions. The 1978 genetic gain trial series yielded valuable and convincing rotation-age evidence of realised gain from radiata pine improvement efforts, which has contributed to the forestry sector's continuing strong support for the breeding programme.

Table 16.1 summarises gains realised at a high-productivity site of the 1978 genetic gain trial series. Although this result is exceptional, there is substantial

**Table 16.1** Comparison for timber volume, percentage high-quality sawlogs, gross revenue per hectare and gain in $NPV/ha of unimproved radiata pine seedlots with both an OP seed orchard seedlot and a good CP cross seedlot, from age 30 measurements in a replicated large-block genetic gain trial established at Mohaka Forest in 1978

| Genetic seed source | Timber volume ($m^3$/ha) | High value log grades (%) | Revenue at age 30 ($) | Value gain per hectare of improved seedlots ($NPV/ha @ 8% interest rate) |
|---|---|---|---|---|
| Unimproved seedlot | 754 | 61 | 67,300 | – |
| OP seed orchard seedlot | 905 | 68 | 82,900 | 1550 |
| Good CP cross | 1165 | 69 | 108,800 | 4130 |

evidence from the subsequent silviculture/breed trials indicating that national tree volume gains/year from the NZ breeding programme are averaging in excess of 1% per year of breeding effort, which is consistent with gains being achieved in a number of other tree and plant crop species (Kimberley et al. 2015).

A number of recent publications have addressed the goal of estimating realised gains based on measurements of these long-term genetic gain trials and the inclusion of these estimates in NZ tree crop models. Genetic multipliers were first developed in the 1990s for modelling improvements in growth rate for individual genotypes (Carson et al. 1999b). Using a similar approach and based on similar evidence for realised gains in wood density and corewood stiffness (Carson et al. 2014), Scion researchers have now developed genetic multipliers for improvements in these traits (Kimberley et al. In Press).

An ideal future approach to realised gain prediction and genetic gain modelling would be to (1) plant the range of genetic stock available for planting in a production forest in progeny trial-like experimental designs across the forest and (2) systematically allocate permanent sample plots of the type used by mensurationists for growth model development and validation to sites, representing the range of available genetic stock. Elements of this approach are already being tried in NZ plantations, in which 'paired plots' of unimproved and improved seedlots are being established within plantation stands and inventory plots are, in turn, being embedded in progeny trials. In the future, although physiological models developed for growth prediction have disappointed in the past, a new hybrid model combining physiological data with empirical measurements is showing promise and may assist gain prediction through an improved understanding of the underlying causes of G × E (Mason et al. 2018).

Another promising approach for improved prediction of genetic gain in plantation forests would be to substitute the age/age correlations used in the best linear unbiased estimation (BLUP) approach commonly used by tree breeders with estimates derived from crop model predictions. More accurate predictions could then be made for a range of silvicultural prescriptions, rather than assuming that realised gain will be the same under all planting conditions.

## 16.16 Major Challenges and Opportunities

Tree breeding programmes are subject to constant challenge, review and revision as new technologies and new thinking are brought to bear. The overall challenge for the NZ radiata pine breeding programme is to increasingly deploy to achieve substantial additional genetic gains/year in key traits, to provide protection against major risks posed by pests and diseases and to retain sufficient genetic diversity to maintain future gains while managing any effects of inbreeding. Genomic selection represents a technology that is close to implementation. Other attractive near-term opportunities are available, including:

Cloning of the breeding population and adoption of clonal deployment as the approach which delivers the maximum genetic gain in a cost-effective way.
New accessions sourced both from other breeding and clonal programmes and from 'mining' of existing plantations and gene editing and gene transfer.

## 16.17 Cloning the Breeding Population and Adoption of Clonal Deployment

Somatic cloning methods offer the potential to clone the breeding population, which could lead to a 'step-change' improvement in delivery of genetic gains. The potential benefits include:

Somatic clones that are tested in the breeding programme as potential forwards selections can be immediately mass-propagated and deployed in plantations through clonal deployment. This substantially reduces the time required for conventional seed propagation and deployment and delivers both higher genetic gain and increased gains/year from the breeding programme (but as yet, only 10–15% of seeds and few families yield somatic clones).

The capability to combat a new disease or other threat would be enhanced. Resistant stocks could be deployed more rapidly because somatic clones in cryostorage can be deployed as bulked or single clones for deployment within 3–5 years of recognition of a crisis. In addition, resistant genotypes can be more quickly established in breeding archives and deployed as seed from seed orchards and/or as cuttings from control-pollinated stoolbeds.

Direct deployment of somatic clones represents the best platform for deploying new genomic technologies. This includes genomic selection, as well as CRISPR-type gene editing and gene transfer methods. Whether in response to a threat or for making conventional breeding gains, improved stock could be delivered to plantations via the above propagation options within a 3–5-year timeframe.

Somatic clones can also be transferred in vitro from NZ to Australia, for testing and deployment in both countries, offering key opportunities for improving both gains and connectivity.

The main technical challenges revolve around the current costs of capturing genotypes into somatic embryogenesis and the low efficiency of the capture process. Only 10–15% of immature seeds cultured will become propagable somatic clonal lines. However, there is no evidence to date for any reduction in genetic diversity as a result of the low efficiency of the capture process, and the costs of clone line capture can be considered manageable when compared to the potential benefits. One of the largest forest owners in NZ has embraced this technology for deployment, and a pilot programme comprising over 1500 cryostored NZ breeding clones has been recently initiated.

## 16.18 New Accessions

Several of the recent changes in breeding strategy have arisen from the realisation that the NZ breeding population is now almost 'closed', in the sense that there are relatively few options for expanding genetic diversity without reverting to making selections in the relatively unimproved native provenances. This, however, represents only one of several unexercised opportunities to access and utilise potentially available genotypes.

New accessions could come from at least four different sources, that is:

Other breeding programmes with radiata pine, including progeny-tested selections from the Forest New South Wales, Southern Tree Breeding Association and Chilean Tree Breeding Association programmes. In each case, there are potential biosecurity barriers to overcome, but these may be resolvable with some creative planning and associated R&D investment.

Accessions from these programmes are likely to provide substantial additional genetic diversity, since they will be largely unrelated to material already in the NZ breeding population.

As mentioned already (under 'Genomic Selection'), two NZ providers of somatic clones currently hold several thousand clonally tested genotypes in their cryostorage facilities, archives and field trials. Although these clones represent progeny and siblings of breeding parents already in the NZ breeding population, their introduction into the breeding population and provider seed orchards would undoubtedly create options for increasing gains/year in both the breeding and deployment populations.

A similar potential exists within the NZ plantation resource, even although this resource is largely comprised of seed orchard progenies derived from the NZ breeding programme. The NZ radiata pine estate covers 1.5 million hectares and has been planted within excess of 1.2 billion individual genotypes, of which in the order of 400 million trees are expected to be present at harvest age. In terms of population genetics, this vast genetic resource must be assumed to contain a very large number of genotypes possessing new gene combinations and mutations, when compared to the relatively tiny sample of genotypes being progressed in the breeding programme. New plus trees identified and sampled from this resource could be screened for their pedigree and performance, for example, using the new GS SNP panel, thereby creating selection options for achieving additional gains and genetic diversity in both breeding and deployment.

As mentioned above, new selections can be made in various existing NZ and Australian trials and stands containing material from the five native provenances of radiata pine. To date this has been an unattractive option, since (with the possible exception of some genotypes from the Guadalupe provenance) these native populations have progressively fallen further behind the NZ breeding population for their performance for the key target traits. Selections within them would likely need to be backcrossed with better parents for several breeding cycles before they could contribute to gains in deployment. However, these native populations offer potential contributions of valuable alleles not present in the breeding population, and such genetic variation could become critically important in combating any new serious threats from pests and diseases. Also, the new genomics tools should make it possible to make the backcrossing process more efficient, such that these new alleles could be brought into play much more rapidly than otherwise (J. McEwan, personal communication).

## 16.19 Genetic Manipulation

There has been a plethora of academic and public debate about the pros and cons of genetic modification (GM), and this is likely to be ongoing. In NZ, the Royal Commission's recommendations in 2001 were favourable to cautious commercial deployment of GM plant stocks. However, since then there has been little practical progress towards achieving commercial applications of GM for radiata pine, despite years of subsequent research that has indicated both the high potential and the low risks associated with this option (Walter et al. 2005).

As was the case in 2001, the threat of negative public perceptions has continued to be the main barrier to implementation of GM methods, combined with an understandable reluctance among the large forest growers to be the first to challenge the *status quo*. Whether perceptions and regulations associated with the new gene editing methods will be any different remains to be tested. It appears clear, however, that these new techniques offer much more accurate manipulation of existing genes, without necessarily incurring the risk and/or fears currently associated with foreign gene transfers.

## 16.20 Conclusions

NZ radiata pine breeders have been highly successful in increasing the productivity of the large forest estate of radiata pine in NZ and parts of Australia. Improved statistical methods, trial designs and decision support tools are making breeding more efficient, while new indirect and remote screening methods show promise of increasing gains/year, particularly in deployed treestocks. New information has enabled simplification of the breeding population structure, and assortative mating will assist achievement of a balance among genetic gain, diversity and inbreeding effects. Opportunities are available for accessing other sources of improved genetic material, thereby enabling an increased emphasis to be placed on selection intensity, leading to increased genetic gains. Cryostorage and retrieval of somatic clones in juvenile form provide exciting options for immediately increasing realised genetic gain through deployment and breeding, as well as being the preferred platform for successful implementation of genomic selection and for potential deployment of improved, gene-edited treestocks. These biotechnology tools, when combined with sound conventional tree breeding practice, may well bring about a step-change improvement in future cycles of NZ tree breeding.

**Acknowledgements** A large number of colleagues at the NZFRI over many years have provided the basis for the NZ radiata pine tree improvement programme. Recent discussions with a number of current tree breeding colleagues have also contributed ideas and analysis, and in particular we thank Luis Apiolaza, Brian Cullis, Rob Woolaston, Shaf van Ballekom, Ruth McConnochie, Fred Burger, (the late) Paul Jefferson, Tony Shelbourne, Phil Wilcox, John McEwan, Dave Evison and Heidi Dungey for their inputs.

## References

Jayawickrama, K.J.S. and M.J. Carson, 2000; A breeding strategy for New Zealand radiata pine. Silvae Genetica 49 (2000), pp. 82–90.

Shelbourne, C. J. A., R. D. Burdon, S. D. Carson, A. Firth and T. G. Vincent. 1986: Development plan for radiata pine breeding. Forest Research Institute, Rotorua, New Zealand. 142pp.

H. S. Dungey, J. T. Brawner, F. Burger, M. Carson, M. Henson, P. Jefferson and A. C. Matheson. 2009. A New Breeding Strategy for *Pinus radiata* in New Zealand and New South Wales. Silvae Genetica 58, 1–2 (2009).

Kinghorn, B.P., Shepherd, R.K. 1999. Mate selection for the tactical implementation of breeding programs. Assoc. Advmt. Anim. Breed. Genet. 13, 130–133, http://www.aaabg.org/livestocklibrary/1999/AB99025.pdf.

Harry X. Wu, Henrik R. Hallingbäck, and Leopoldo Sánchez. 2016. Performance of Seven Tree Breeding Strategies Under Conditions of Inbreeding Depression, Genes/Genomes/Genetics Volume 6, pages 529–540| March 2016.

Borralho, N. M. G., and G. W. Dutkowski, 1998 Comparison of rolling front and discrete generation breeding strategies for trees. Can. J. Res. 28: 987–993.

Alzamora, R.M., (2010) Valuing breeding traits for appearance and structural timber in radiata pine. PhD thesis. School of Forestry, University of Canterbury, 2010.

Ivkovic M., H. Wu and S. Kumar. 2010. Development of breeding objectives for structural and appearance grade products: Bioeconomic modelling as a method for determining economic weights for optimal multiple trait selection. Silvae Geneticae, 59, 2–3 (2010).

Evison D and L. Apiolaza. 2014. Incorporating economic weights into radiata pine breeding selection decisions. Can J For Res 45:135–140.

Luis A. Apiolaza and Dorian J. Garrick. 2001. Breeding objectives for three silvicultural regimes of radiata pine, Can. J. For. Res. 31: 654–662 (2001).

King, J. N. and Burdon, R.D. 1991. Time trends in inheritance and projected efficiencies of early selection in a large 17-year-old progeny test of *Pinus radiata*. Can. J. For. Res. 21: 1200–1207.

Li Li and Harry X Wu. 2005. Efficiency and early selection for rotation-aged growth and wood density in *Pinus radiata*. Can.J. For. Res. 35 2015–2029.

S. Kumar & J. Lee. 2002 Age-Age correlations and early selection for end-of-rotation wood density in radiata pine. Forest Genetics 9 (4) 323–330.

Shelbourne, C. J. A., S. Kumar, R. D. Burdon, L. D. Gea and H. S. Dungey. 2007: Deterministic Simulation of Gains for Seedling and Cloned main and Elite Breeding Populations of *Pinus radiata* and Implications for Strategy. Silvae Genetica 56: 253–300.

Alison B. Smith and Brian R. Cullis (In Press) Plant breeding selection tools built on factor analytic mixed models for multi-environment trial data. TAG.

Cullis, B.R., Jefferson, P, Thompson, R, Smith, A.B. 2014. Factor analytic and reduced animal models for the investigation of additive genotype by environment interaction in outcrossing plant species with application to a *Pinus radiata* breeding program. Theoretical and Applied Genetics 127:2193–2210.

Carson, M.J. and C. S. Inglis, 1988. Genotype and location effects on internode length of *Pinus radiata* in New Zealand. New Zealand Journal of Forestry Science 18 (3): 267–79 (1988).

Apiolaza, L.A., (2009). Very early selection for solid wood quality: screening for early winners. Ann For Sci. 66 (2009) 601 Available online at: c_ INRA, EDP Sciences, 2009 www.afs-journal.org DOI: https://doi.org/10.1051/forest/2009047.

Apiolaza, L.A., John C.F. Walker, Hema Nair, Brian Butterfield, 2008 Very early screening of wood quality for radiata pine: pushing the envelope. Proceedings of the 51st International Convention of Society of Wood Science and Technology November 10-12, 2008 Concepción, CHILE.

Carson, S. D. 1989. Selecting radiata pine for resistance to Dothistroma needle blight. N. Z. J. For. Sci. 19 (1):3–21. (Reprint No. 2233).

Carson, S. D., and M. J. Carson. 1989. Breeding for resistance in forest trees - a quantitative genetic approach. Ann. Rev. Phytopathology 27:373–95. (Reprint No. 2226).

Heidi S Dungey, Nari M. Williams, Charlie B. Low and Graham T. Stovold, 2014. First evidence of genetic-based tolerance to red needle cast caused by *Phytophthora pluvialis* in radiata pine. N Z J F. Science, December, 2014.

Matheson, A.C.; Devey, M.E.; Gordon, T.R.; Balocchi, C.; Carson, M.J.; Werner, W. 2006. The genetics of response to inoculation by pine pitch canker (*Fusarium circinatum* Nirenberg and O'Donnell) infection by seedlings of radiata pine (*Pinus radiata* D. Don). Australian Forestry. 69 (2): 101–106.

Isik F, Bartholome J, Farjat A, Chancerel E, Raffin A, Sanchez L, Plomion C, Bouffier L. 2016: Genomic selection in maritime pine. Plant Sci 242:108–119.

Resende M.D.V., Resende M.F.R., Sansaloni C.P., Petroli C.D., Missiaggia A.A., Aguiar A.M., Abad J.M.,Takahashi E.K., Rosado A.M., Faria D.A., Pappas G.J., Kilian A, Grattapaglia D. 2012 Genomic selection for growth and wood quality in Eucalyptus: capturing the missing heritability and accelerating breeding for complex traits in forest trees. New Phytol 194:116–128.

Dario Grattapaglia. 2017. Status and Perspectives of Genomic Selection in forest Tree Breeding. Chapter 9 InGenomic Selection for Crop Improvement New Molecular Breeding Strategies for Crop Improvement. Eds. Rajeev K. Varshney, Manish Roorkiwal, Mark E. Sorrells. Springer.

M.B. Powell, T.A. McRae, H.X. Wu, G.W. Dutkowski, D.J. Pilbeam (2004) Breeding Strategy for *Pinus radiata* in Australia. *2004 IUFRO joint conference of division 2, forest genetics and tree breeding in the age of genomics: progress and future*, Charleston, South Carolina, USA – 1-5 November 2004.

Carson, S. D., M. O. Kimberley, J. D. Hayes, M. J. Carson. 1999a. The effect of silviculture on genetic gain in growth of *Pinus radiata* at one-third rotation. Canadian Journal of Forestry Research 28 (2):248–258.

Carson, S.D., D. J. Cown, R. McKinley, and J. R. Moore. 2014. Effects of site, silviculture and seedlot on wood density and estimated wood stiffness in radiata pine at mid-rotation. New Zealand Journal of Forestry Science 44 (1).

Carson, S. D., O. Garcia, and J. D. Hayes. 1999b. Realised gain and prediction of yield with improved *Pinus radiata* in New Zealand. Forest Science 45 (2):186–200.

Mark O. Kimberley, John R. Moore, Heidi S. Dungey. 2015: Quantification of realised genetic gain in radiata pine and its incorporation into growth and yield modelling systems. *Canadian Journal of Forest Research*, 45 (12): 1676–1687, https://doi.org/10.1139/cjfr-2015-0191.

Mark Kimberley, John Moore and Heidi Dungey (In Press) Quantification of realised genetic gain for wood stiffness in New Zealand radiata pine.

Mason E.G., Holmström E. and Nilsson U. 2018. Using hybrid physiological/mensurational modelling to predict site index of *Pinus sylvestris* L. in Sweden: a pilot study. Scandinavian Journal of Forest Research 33 (2): 147–154. https://doi.org/10.1080/02827581.2017.1348539.

Walter, C., Carson, M. and Carson, S.D. 2005: Biotechnology applications to conifer plantation forestry. In Walter, C. and Carson, M. (eds.) Forest Biotechnology for the 21st Century. Research Signpost. Trivandrum, Kevali, India.

Li Y, and Dungey H.S. (2018) Expected benefit of genomic selection over forward selection in conifer breeding and deployment. PLOS ONE 13(12): e0208232. https://doi.org/10.1371/journal.pone.0208232

# Appendix

## List of all provenance trials planted in NZ 1954-1980

**Eastern USA**

***Pinus banksiana*** 1955/56 six seedlots, Kaingaroa, Naseby. 2 plots/site. 225 trees/plot (Kaingaroa), 360-400 (Naseby)

***Pinus resinosa*** 1956 seven seedlots, Kaingaroa, Karioi, Berwick, Naseby 2-3 plots/provenance/site, no data on trees/plot

***P. elliottii var. elliottii*** 1956 eight provenances, 17 OP seedlots, Waitangi, 1-3 plots/seedlot, 144 trees/plot

***P. taeda*** 1955 18 seedlots (13 provenances and bulks), Waitangi, Rotoehu, 1-3 plots/seedlot/site, 196 trees/plot

***P. strobus*** 1970 77 seedlots, Gwavas, Rotoehu, Golden Downs, 5 reps., 10-tree row plots, Sets in Reps. design with ca. 20 seedlots per set.

**Europe and Asia**

***P. pinaster*** 1955 47 seedlots, 9 sites, Woodhill, Waitangi (RCB), Rotoehu, Berwick, Puhipuhi, Berwick, Naseby. 2-3 plots/seedlot at Woodhill, Waitangi, rest 1 plot/seedlot, 196 trees per plot at RCB sites, others 100 trees/plot

***P. sylvestris*** 1956 Five seedlots 3 sites, Kaingaroa, Berwick, Naseby, 1 plot/provenance/site, 100, 121, 190 trees per plot respectively

***Larix decidua*** 1959 27 native, 10 exotic seedlots, Kaingaroa x 2.(1-2 plots/seedlot, 100 trees/plot), Patunamu, Gwavas, Golden Downs, Hanmer, Berwick, Rankleburn, Naseby, Whaka (1-3 plots/seedlot, 121-144 trees/plot)

***Larix leptolepis*** 1957 20 provenances, Kaingaroa x 2, 1-3 plots/seedlot, Patunamu, Gwavas. 3 plots/seedlot, Golden Downs, Hanmer, Berwick, Naseby, Whaka 1-3 plots per seedlot, 100-196 trees/plot

***P. nigra*** 1956-58 39 seedlots (native populations), 5 ex European plantations, 4 NZ at Whaka, Gwavas, Golden Downs, Karioi, all one plot/provenance. Karioi, Golden Downs, Ngaumu, 3 plots/provenance. 100 trees plot except 144 at Golden Downs and Karioi.



C. J. A. Shelbourne, M. Carson, *Tree Breeding and Genetics in New Zealand*,
https://doi.org/10.1007/978-3-030-18460-5

**Western North America**

***P. attenuata*** 1961 49 seedlots, from southern Oregon to south California, including 2 hybrids with radiata. Designs: 3-4 reps RCB of 16 trees/plot/site, and 7x7 lattice with 2 rows of 6 or 8 trees/plot. Only 11 seedlots with 2-6 OP seedlots/provenance. Sites, Whaka, Kaingaroa, Karioi, Naseby

***P. muricata*** 1972 Rotoehu only (Bannister) 16 provenances, 15 OP families/provenance, 236 families, probably 20 trees per family max.

***P. muricata*** 1973 Trinidad, Mendocino, Sonoma, 2 NZ blue (Mendo.), Marin, NZ green (Marin). 10 sites, Kaingaroa (x2), Karioi, Kaweka, Golden Downs, Mahinapua, Naseby, Berwick, all 10-tree rows, RCB, 5-10 reps.

***P. contorta*** 1958-61 14 seedlots by Egon Larsen, 11 by USFS, 3 NZ. 7 coastal provenances, 13 murrayana, 5 latifolia, 2 intermediates. 15 sites, 3-4 plots/prov./site. No information on trees/plot but assume 100+. Esk and Kaingaroa 25 plots/provenance, 2 trees/plot.

***P. ponderosa*** 1954 Five seedlots all within 50 miles of lat. 38° 30'. 1 plot/seedlot/site at 9 sites, 169-324 trees/plot, 3 plots/seedlot at Gwavas, 2-4 Golden Downs, 2 Berwick. 170-324 trees/plot.

***P. ponderosa*** 1960-61 8 sites 1960, 8 in '61, 1 in '62. 64 seedlots. RCBs, 14 sites with 3 plots/seedlot/site, 13 sites with 4 plots, Esk 1 plot/rep, 32 trees/plot.

***Picea sitchensis*** 1972 RCB 5-10 reps/site, 10 tree row plots, 8 sites (4 per island), Hunua, Karioi, Slopedown, Whaka

***Picea sitchensis*** 1956 4 seedlots, Big Lagoon, Caspar (Cal.), Waldport, Ore.,Vancouver Is., Naselle, Wa. 1-3 plots/site., 17 sites, 11 of 17 abandoned by age 24.

***Pseudotsuga menziesii*** 1957 mostly commercial lots, Darrington, north Wa. to Siskyou, south Ore. also Vancouver Is., No coastal provenances. 7 sites 100-196 trees/plot, Rotoehu (1/2 121), Kaingaroa (3 196), Kaingaroa (1 99), Patanamu (1 121), Golden Downs (3 ?), Hanmer (2/3 196), Berwick (3 144)

***Pseudotsuga menziesii*** 1959 20 sites, 121-144 trees/plot, 1-3 plots/seedlot/site, mostly 1 plot. Sites with 2 or 3 plots/provenance: Glenbervie, Rapanui (NZFP), Kaingaroa Cpt. 1149, Gwavas, Golden Downs, Hanmer, Rankleburn.

I haven't been able to access any information on ***Tsuga heterophylla*** and ***Thuja plicata***, ***Tsuga*** definitely planted at Hanmer. Both important species in Western forests. ***Populus trichocarpa*** and ***Alnus rubra*** are also both important.

**Australia**

***Eucalyptus regnans*** 1977 two sites, Wiltsdown (Tokoroa) and Kaingaroa. 144 OP families in 36 provenances 36 reps, 'sets in reps' design (sort of!) Sets A, B, C, D at Wiltsdown Set D at Kaingaroa Set A Tasmania, Set B Victoria, Set C NZ, Set D mixed.

***E. fastigata*** 1979 Kinleith, Kaingaroa 126 seedlots total. 115 OP families 11 composites, from NSW and Victoria, 1 South Africa, 15 NZ stands, 51 NZ OP seedlots. Field trials 3 sets of 42 seedlots 36 reps STPs at Kinleith, 42 reps STPs at Kaingaroa.

***E. saligna*** 1976 43 native OP families (as 18 provenances), 55 NZ families, 4 sites, STPs, ?? no. of sets. Includes *E. botryoides)* (1), *E. grandis* (3), *E. deanei* (2), *E. dunnii* (1)

***E. delegatensis*** 1977 38 native OP families, 52 NZ, RCB, 27 reps of STPs/site, 2 sites

***Eucalyptus* species trial** Longwoods, 1977 RCB 3 reps of Main plots (of species), 15, 30 or 45 trees. Provenances were STPs within Main species plots. Average 9 trees/prov., 5 provs./species.

***E. nitens*** 1979 80 OP Victorian families plus a few NSW families. No data. ca. 4 sites inc. Longwoods, Tongariro, Kaingaroa, STPs, sets in reps

***E. nitens*** 1990 Second gen. selections. 300 OP families, two sites STPs and 1 site forwards selection row plots

*Pinus radiata*

***Pinus radiata*** 1955 29 OP families 3 reps of 20 tree row plots (some select OPs as in 850 55) 20-tree-row plots, 3 sites, Gwavas, Golden Downs, Berwick, assessed 1967 age12. First estimates of genetic parameters from OP progenies, range of narrow sense $h^2$. Gave confidence for use of OP in '268' programme, starting in 1968/9.

***P. radiata*** 1955 Two provenances of radiata, Monterey, Ano Nuevo. 2 sites, Golden Downs, Gwavas. Big plots 100-200 trees/plot 2 or 3 plots/prov/site.

***P. radiata*** 1957 Assorted CP crosses including many with '850'-19, '850'-55 (and '850'-20) as 'testers', plus a Kaingaroa bulk unselected lot. These provided the first estimates of genetic gains from radiata in NZ. 20 tree-row-plots, 3 reps in RCB. Total number of crosses (not including fastigiate clone 20) about 20.

***P. radiata*** 1964-66 Genetic survey of *P. radiata* ca. 350 OP families (5 native and 2 NZ populations with 50 OPs each. STPs with interlocking block design to facilitate thinning (Libby and Cockerham).

***P. radiata*** 1968 '266' clonal tests, 216 clones selected in 1964 in 5- and 6-year-old stands (6 & 7 years from seed). Two sites, Cpt. 1350 Kaingaroa and Whaka forest. First STP design, 3 reps/site 3-6 ramets of each clone/rep. Laid out as Lattice.

***P. radiata*** 1968 progeny test of 14 OP families ex seed orchard and CP families of most of these forming factorial of 8 '850' series clones from first orchard. 1 site. STPs 3-5 seedlings/family/rep. About 10 reps. 4 x 4 factorial results published by Wilcox, Shelbourne and Firth.

***P. radiata*** 1969 progeny test of 588 OP families of '268' series. 3 sites, Kaingaroa Cpt. 1350, Waimihia, Gwavas. Sets in reps. 16 sets of 38 families, '850' orchard lot, bulk unselected lot, 10-tree row-plots (Set 16 is a half-set).

***P. strobus*** 1970 77 mainly provenance seedlots, Gwavas, Rotoehu, Golden Downs, sets in reps, 20 seedlots/set, 5 reps, 10-tree plots.

***P. muricata*** 1972 (Bannister) RCB 15 fams/provenance, 236 OP families, ca. 18 provenances. including NZ blue and green (Mendocino and Marin). 1 site Rotoehu, Trinidad, Mendocino, Sonoma, Marin, Monterey, San Luis Obispo, Santa Rosa, Santa Barbara 16 provenances plus Cedros radiata, small row plots RCB.

***P. muricata*** (Shelbourne) 1973 2 sites, Cpt. 1038 and 885 Kaingaroa, assorted Sonoma plus others. 10 site countrywide trials with Trinidad, Mendocino, Sonoma, Marin, 2 x NZ blue and 1 x NZ green (Marin) all RCB 10-tree row plots 5-10 reps.

***P. radiata*** 1972 '268' OP tests Golden Downs, Woodhill, Otago Coast, (Ashley failed), 220 of 588 '268' OP families plus 80 SI OP families, sets in reps, 30 families/set, 10-tree row plots.

***P. radiata*** 1975 Polycross 6 sites Kaingaroa (Western boundary Cpt. 327), Berwick, Golden Downs, 4 sets in reps, STPs but multiple trees (2-5) per rep., 100 families 25-30 per set

***P. radiata*** 1975 Disconnected Diallel '850' GE study, 11 sites, 25 parents, 5 half diallels, one 5x5 full diallel, sets in reps, diallels coincident with sets, 2-5 seedlings per cross per rep, 6 reps. Sites from Woodhill south to Berwick.

***P. radiata*** 1978 Genetic Gain trials. Type A Row plot of ca. 10 entries, RCB 8-10-tree row plots, ca. 10 sites.

Genetic Gain Type B large plot trials B1 sawlog regime, pruning and thinning to ca. 300 s/ha C. B2 pulpwood regime, no thinning, no pruning. Different initial spacing. RCBs, number of reps 6 Kaingaroa, Mohaka, Woodhill, Golden Downs, Rankleburn

***P. radiata*** 1980 '875' Disconnected Diallel second generation test STPs ca. 20 5 x 5 disc. half diallels, ca. 100 '875' parents, 2 or 3 sites, Kaingaroa Cpt. 327 and Tasman seed orchard, Kawerau

***P. radiata*** 1980 Full-sib selection blocks '268', 1 or 2 49 tree blocks for within-family selection from high GCA crosses among '268's. On Rotorua-Whakatane hwy. Plus 2 sites with STP trials

***P. radiata*** '880' second generation OP families 1982 4 sites, Moerewa, Rotoehu, ?? STPs sets in reps

***P. radiata*** '885' and '887's Additional BP, first gen., STPs, sets in reps design, OP tests (as at Kaingaroa Cpt. 327 but where else?)

***P. radiata*** Native populations collected by Firth and Eldridge in 1978. OP progenies bulked to make single sub-population lots, sub-populations bulked for field gene resources. Plot size ? ca. 49 planted throughout NZ.

## List of Propps and Sidney Reports

C.J.A. Shelbourne 1965: Current misconceptions about field designs for progeny testing. PhD Proposition, NC State University, School of Forestry.

A.D. Lindsay (1936?) Bishop pine (*Pinus muricata* D. Don) in its native habitat. Comm. For. Bur. Aust. Bull. No. 11.

J.M. Fielding 1961 Provenances of Monterey and Bishop pines. For. Tim. Bur. Aust. No 38.

**Douglas-Fir**

R25/6924 T.G.Vincent 197I: Selection for wood density in Douglas-fir

IR28/6927 T.G. Vincent 1971: Possible provenance-planting year differences in Kaingaroa Douglas-fir.

10/6085 D.V. Birt and T.G. Vincent 1971: Selection for wood density in Douglas-fir.

33/6930 A. Firth 1972: Provenance variation in Douglas-fir-Waitangi.

58/6963 M.D. Wilcox and T.G. Vincent 1974: Third year results of Douglas-fir seed source test.

60/6965 M.D. Wilcox and T.G. Vincent 1974: Comparison of Californian, Mexican and Kaingaroa Douglas-fir at the Long Mile trial area, Rotorua.

58/6963 M.D. Wilcox and T.G. Vincent 1974: Third year results of Douglas-fir seed source test.

114/7021 M.D. Wilcox 1976: First assessment of Douglas-fir open-pollinated progeny tests at age four years from planting.

13/7020 J.T. Miller 1976: Final assessment of provenance trial in *Ps. menziesii* at Cartwheel Hut, Rankleburn forest.

114/7021 M.D. Wilcox 1976: First assessment of Douglas-fir open-pollinated progeny tests at age four years from planting.

162/7073 J.T. Miller and M.D. Wilcox 1976: Shelterbelt provenance trial in Douglas-fir in Canterbury.

531/8215 M.D. Wilcox, I.J. Thulin and J.T. Miller 1987: Growth comparisons of commercial Douglas-fir seedlots from native and NZ sources.

6074/15561 C.B. Low and M.A. Miller 1993: (1998) Douglas-fir seed collection in California and Oregon 1993.

3867/2794 J.T. Miller and C.J.A. Shelbourne 1994: Douglas-fir in Southland: a report on a visit to the forests of Earnslaw One Ltd.

4879/14108 T.G. Vincent 1995: Inspection of potential Douglas-fir seed orchard sites in central Otago and western Southland. A report for Earnslaw One Ltd.

5329/14754 C.B. Low 1996: Flushing times and early height growth from USA and NZ Douglas-fir seed source test ( from Low and Miller, M.A. seed collection 1993)

5858/15466 G.T. Stovold 1997: Establishment report for 1996 Douglas-fir open-pollinated progeny tests on three sites.

5851/15454 G.T. Stovold 1997: Establishment report for Douglas-fir seed source trials at seven sites.

IR 212 PROPP 686 J.T. Miller 2003: Species trial of *P. attenuata x P. radiata* hybrids.

Sydney 36317 H. Dungey and C.B. Low : Development of hybrids with *P. radiata.*

**Cypress**

1291 D.C. Maika 1999: Trial establishment of 2nd. generation progeny test of *C. lusitanica.*

30806 J.T. Miller 2000: Cypress strategy.

30424 C.B. Low and D.C. Maika 2000: Trial establishment report of 2nd. generation progeny test of *C. macrocarpa*

32928 B.V. Geard 2001: Comparison of the growth rate, form, branching, health and wood characteristics of clones and seedlings of *Cupressus spp.*

35199 G.T. Stovold and D.G. Holden 2002: *C. lusitanica* clonal trial establishment.

36398 G.T. Stovold and D.G. Holden 2004: Establishment of *C. lusitanica* clonal trial.

36406 G.T. Stovold and D.G. Holden 2004: Establishment of *C. macrocarpa* clonal trial 2003.

39587 C.B. Low and S.J. Gatenby 2005: Assessment results of the 1985 cypress trial at 3 locations.

37654 T.R. Chandrasekhar, S. Kumar and C.B. Low 2006: Preliminary estimates of genetic parameters of early growth in a clones-in-families test of *C. lusitanica.*

39583 C.B. Low 2006: Age 18 assessment of cypress hybrid clonal trial at Gwavas.

41081 K.R.Fleet, G.T. Stovold and C.B. Low 2007: Establishment report for 2006 *C. lusitanica* progeny trials.

45320 C.B. Low, M.A. Miller and D. Grogan 2009: Establishment report for cypress hybrid trials.

**Radiata Pine**

4. G.B. Sweet and M.P. Bollman 1962. Provenance trial in *Pinus sylvestris*: Establishment report

40/6939 T.G. Vincent and M.D. Wilcox 1975: Frost damage at Matea, Kaingaroa: a comparison of *P. muricata, P. radiata* and *P. contorta.*

13-6912 C.J.A. Shelbourne 1970: Background and progress report on international gene pool and seedling orchard evaluation plantation for 2nd generation selection of PR.

47-6952 M.D. Wilcox and D.V. Birt 1974: Clonal variation in wood density in 5-year old rooted cuttings of PR in Cpt. 32 Whaka forest.

51-6956 M.D. Wilcox, A. Firth and C.B. Low 1974: Performance at 5 years of block plantings of full sib families of PR seed orchard clones ('850' series).

35-880 C.J.A. Shelbourne and O. Mohrdiek 1982: Assessment of polycross progeny tests at 6 sites age 6 years from planting.

110-802 M.J. Carson, C.J.A. Shelbourne and C.B. Low 1982: Final assessment of uninodal progeny trials.

104-808 M.D. Wilcox 1982: Selfs and outcrosses progeny test of radiata pine.

242-657 R.D. Burdon, C.B. Low and A. Firth 1983: Results to age 10 from NC2 among '850' PR clones at Waimihia and Goudies block.

212-686 J.T. Miller 1983: Species trial of *P. attenuata* hybrids.

172-729 C.J.A. Shelbourne 1983: Interracial hybrids between NZ RP and Cedros and Guadalupe island populations at ages 14 and 8.5 years respectively.

269-626 M.D. Wilcox 1983: Three reports on 266' clonal tests.

159-742 C.J.A. Shelbourne 1983: Comparison between PR families and orchard/ provenance groups in the International Gene Pool Experiment at age 6.5 years.

242-657 R.D. Burdon, C.B. Low and A. Firth 1983: Results to age 10 from NC2 among '850' PR clones at Waimihia and Goudies block.

684-107 C.J.A. Shelbourne 1984: Combined ranking of all NI '850' seed orchard clones from index rankings in progeny tests.

454-31 J.T. Miller 1984: Assessment of 16-year-old *P. attenuata* provenance trial at Naseby forest.

478-338 C.J.A. Shelbourne, A. Firth and C.B. Low 1984: Genetic gains at age 5 years from 3 regional seed orchards of '850' series clones.

441-359 S.D. Carson 1984: A preliminary plan for breeding RP, resistant to *Dothistroma pini*, and needle-cast associated with *Nemacyclus minor*.

561-231 P.J. De la Mare 1984: Eight-year assessment of '870', '871', '873' p. 1978 clonal tests at Kaingaroa.

728-4293 M.D. Wilcox 1985: 6 year assessment of 870-871 clonal tests (p. 1976) at Kaingaroa.

944-6520 C.J.A. Shelbourne and C.B. Low 1985: Reselection of '850' clones based on growth and form of polycross progenies at 6 sites, age 9 years.

1718-8551 G.T. Stovold 1987; Measurement of genetic gain in volume/ha. in 18-year-old full sib family blocks of '850' series PR.

1952-9064 G.R. Johnson and J.N. King 1988: Age 7 assessment of '875' series diallel, progeny test.

2138-9421 G.R. Johnson 1989: Report on the 1988/89 growth and form assessments of the '880' (OP) tests of Moerewa and Thorpe road.

1988: P.P. Cotterill, C. Dean, J. Cameron and M. Brindbergs 1989: Nucleus breeding: A new strategy for rapid improvement under clonal forestry. In: Proc. IUFRO meeting on breeding tropical trees. Pattaya, Thailand. Nov,-Dec. 1988.

2150-9138 C.J.A. Shelbourne and C.B. Low 1989: Selecting the best '268', '668' and '767'clones from OP progeny tests at Woodhill and Otago Coast forests at age 13 years.

2282-9750 G.R. Johnson, P.W. Volker and L.A. Pederick 1989: An examination of correlations between NZ and Aus. sites for breeding populations.

2283-9751 G.R. Johnson and J.N. King 1989: NZ's radiata breeding programme: the next stage.

R. 2226. S.D. Carson and M.D. Carson 1989: Breeding for resistance in forest trees-a quantitative genetic approach. Annu. Rev. Phytopathol. 27: 373-395.

2577-10357 R.D. Burdon 1990: Genetic parameter estimates on 30 top-ranked'268; PR progenies.

2601-10337 R.D. Burdon 1990: Breeding for wood properties in RP: a problem analysis.

22418-10029 R.D. Burdon 1990: Position Statement: Clonal forestry in PR. (Background paper for Clonal Forestry Symposium, FRI 21.11.89).

3806-12711 C.J.A. Shelbourne 1990: A tale of four Coops. A personal overview.

3807-12712 L.D. Gea, S.D. Carson and P.A. Jefferson 1994: Prediction of parental values for Dothistroma resistance over sites and series.

4261-13278 R.D. Burdon, C.B. Low and M.A. Miller 1994: Countrywide provenance trials of PR at 22 sites. Results of first major assessment at 5-10 years.

3807-12712 L.D. Gea, S.D. Carson and P.A. Jefferson 1994: Prediction of parental values for Dothistroma resistance over sites and series.

4261-13278 R.D. Burdon, C.B. Low and M.A. Miller 1994: Countrywide provenance trials of PR at 22 sites. Results of first major assessment at 5-10 years.

4578-13714 C.J.A. Shelbourne 1995: Future opportunities of Carter Holt Harvey in RP breeding: clonal forestry for end product value.

4462-13545 R.D. Burdon 1995: Preliminary status report on biodiversity in NZ plantation forests.

4676-13835 C.T Sorensson and C.B. Low 1995: Realised genetic gains of RP seed orchard seedlots at age 12 for diameter, tree form and wood density at 14 sites in NZ.

6069-15852 C.J.A. Shelbourne 1995: The future development of tree breeding in Chile.

R. 2992 R.D. Burdon 1995: Future directions in tree breeding: some questions of what we should seek, and how to manage the genetic resource. CTIA/WFGA meeeting, Victoria, BC August 1995.

6064-15846 C.T. Sorensson, D.J. Cown, R.G. Ridoutt and Tan Xin 1998: The significance of wood quality in tree breeding: a case study of RP in NZ.

S.D. Verryn and C.L. Snedden 1998: G-ASSIST version 3.0. A deterministic tool for genetic gains prediction. A user's manual. Water, Environment and Forestry Technology CSIR, Pretoria, South Africa

20932 R.D. Burdon and C.T. Sorensson 1999: Wood properties of RP: update on genetic parameter information and prospects for improvement.

21166 L.D. Gea 1999: Best linear unbiased prediction of DBH breeding values over all sites and series.

L.D. Gea 1999: Best linear prediction of branch cluster frequency.

21269 L.D. Gea 1999: Best linear prediction of straightness; incorporating parent clones of 668, 767, 883, 885, and 887 series.

21865 K.J.S. Jayawickrama 1999: Selecting superior RP families from the NSW progeny trials: results of the 1985, '86 and '87 trials.

20932 R.D. Burdon and C.T. Sorensson 1999: Wood properties of RP: update on genetic parameter information and prospects for improvement.

L.D. Gea 1999: Best linear prediction of branch cluster frequency.

21866 J.R. Lee, L.D. Gea, S.D. Carson 1999: Genetic gains trial to measure reduction of Dothi infection with CP.

31888 K.J.S. Jayawickrama and C.B. Low 2000: Selecting RP clones with fast growth, good form, long internodes and good grain spirality.

30632 K.J.S. Jayawickrama 2000: Genetic parameter estimates for RP in NZ and NSW; a synthesis of results.

31279 T.G. Vincent 2000: Gains from using genetically-improved RP.

30740 L.D. Gea 2000: Best linear unbiased prediction of spiral grain across sites and series.

31751 C.B. Low and M.A. Miller 2001: Assessment of NZ interpopulation Cedros hybrids at age 31.

32013 H.S. Dungey 2001: Genetic parameters and predictions for the advanced generation Dothistroma resistant population age 5.
32673 S. Kumar 2001: Reciprocal and maternal effects in various growth and form traits in 6- year-old RP.
32672 S. Kumar and J.R. Lee 2001: Evaluation of various tools for stiffness measurement in '887' trial in Kinleith Forest.
33243 H.S. Dungey 2002: Early selection using farm field experiment design: results from OP '880' study.
36317 H.S. Dungey and C.B. Low 2004: Development of hybrids with PR.
H.S. Dungey, J.T. Brawner, F. Burger, M.J. Carson, M. Henson, P.A. Jefferson and A.C. Matheson 2005: A new breeding strategy for the Radiata Pine Breeding Consortium: summary and outcomes of the Noosa workshop. RPBC Report?
C.J.A. Shelbourne, R.D. Burdon, H.S. Dungey, S. Kumar, L.D. Gea 2005: Comparison of genetic gains from seedling, versus cloned main and elite breeding populations and associated orchards through deterministic simulation. RPBC Report ?(see 2007 Silvae Genetica).
38627 R.D. Burdon 2005: The RPBC breeding strategy: profit chasing and risk management.
39461 H.S. Dungey 2006: Revisiting the elite-crossing and selection. RPBC/Ensis report? Brainstorming workshop 28th February 2006.
37297 T.G. Vincent 2006: Review of genetic resources of introduced tree species other than PR in NZ.
39193 H.S. Dungey 2006: Analysis of growth and form traits at Kaingaroa Cpt. 60 and Manawahe Cpt. 35.
P.P. Cotterill, C. Dean, J. Cameron and M. Brindbergs 1989: Nucleus breeding: A new strategy for rapid improvement under clonal forestry. In: Proc. IUFRO meeting on breeding tropical trees. Pattaya, Thailand. Nov, Dec. 1988.

**Table 1** Areas from National Forest Survey of pure and mixed combined exotics (acres)

| *Species. Pseudotsuga menziesii* | Area | Percentage |
|---|---|---|
| *P. ponderosa* | 86,179 | 39.9 |
| *P. nigra* | 67,805 | 31.4 |
| *P. contorta* | 26,401 | 12.2 |
| *P. pinaster* | 11,452 | 5.3 |
| *P. muricata* | 6,670 | 3.1 |
| *P. elliottii* | 4,350 | 2.0 |
| *P. strobus* | 3,232 | 1.5 |
| *P. taeda* | 3,105 | 1.5 |
| *P. patula* | 2,963 | 1.4 |
| *P. palustris* | 2,462 | 1.1 |
| *P. sylvestris* | 360 | 0.2 |
| *P. echinata* | 358 | 0.2 |
| *P. lambertiana* | 167 | 0.1 |
| *P. banksiana* | 132 | 0.1 |
| *P. canariensis* | 57 | |
| *P. monticola* | 49 | |
| *P. attenuata* | 12 | |
| *P. pseudostrobus* | 10 | |
| *P. tenuifolia* | 6 | |
| *P. durangensis* | 1 | |
| *P. montezumae* | 1 | |
| **Totals** | **215,772** | |

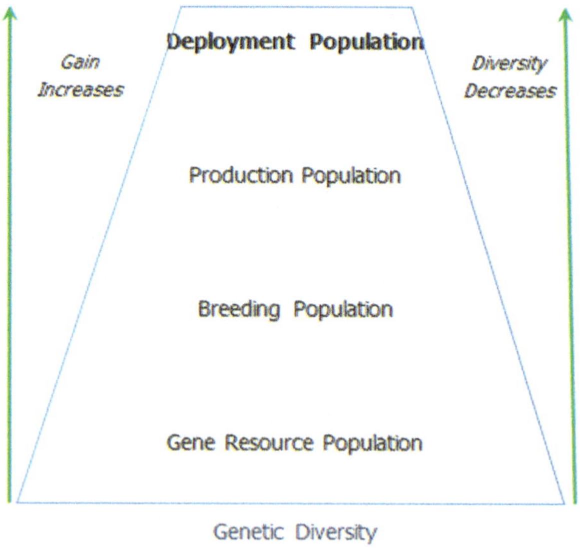


**Fig. 1** The Tree Improvement Triangle (refer Chap. 16)

# References

## Note: the first number in each entry indicates the Production Forestry Division (Branch) report number

## Species and Provenance Testing


2. I.J. Thulin 1962a: Provenance trials of *Pinus banksiana*: establishment report
7 G.B. Sweet 1962a: A provenance trial of *Pinus banksiana:* first assessment report.
8. G.B. Sweet 1962b: A provenance trial in *Pinus resinosa*: first assessment report.
44. C.J.A. Shelbourne 1970: Growth and morphological properties in a combined provenance and progeny test of slash pine *P. elliottii var. elliottii* Engelm.
54. C.J.A. Shelbourne 1971a: Provenance variation in growth rate and other characters in 13-year-old loblolly pine *(P. taeda* L.) in New Zealand.
PR 510 PROPPS 0300 C.J.A. Shelbourne 1984: Choosing the best provenance of Loblolly pine ( *P. taeda* L.) for New Zealand.
IR PROPPS Chen Jianzin 1989 Provenance selection of *Pinus strobus* in New Zealand.
3. G.B. Sweet 1962c Provenance trials of *Pinus pinaster*: first assessment report.
6. G.B. Sweet 1962d. Provenance trial in *Pinus sylvestris*: First Assessment report.
5. I.J. Thulin 1962b Provenance trials of *Larix decidua and L. leptolepis*: Establishment Report.
17. Miller, J.T. and Fairburn, H.S. 1963. Provenance trials in *Larix decidua* and *Larix leptolepis*. First assessment
11. J.T. Miller 1962. The problems of forest tree improvement in larch in New Zealand. Problem analysis.
28. J.T. Miller 1965a. Provenance trial in *Larix leptolepis*: assessment 1964.
29. J.T. Miller 1965b Provenance trial in *Larix decidua*. Second assessment report. Age 6 years
64. J.T. Miller 1973 An extraction thinning in Larch provenance trials.
16. J.T. Miller 1964 Provenance trial of *Pinus nigra*: establishment report
30. J.T. Miller 1965c Provenance trial in *Pinus nigra*. First assessment report
68. M.D. Wilcox and J.T. Miller 1974 *Pinus nigra* provenance variation and selection in New Zealand (See same title Silvae Genetica 24 (5-6): 132-140).
D.J. Cown 1974: Physical properties of Corsican pine grown in NZ. NZJFS (4
15. I.J. Thulin 1963a Provenance trials of *Pinus attenuata*. Establishment report
M.P. Bollman 1966a: The height growth of a provenance trial in *Pinus attenuata* in Kaingaroa Forest Cpt 905.


 
C. J. A. Shelbourne, M. Carson, *Tree Breeding and Genetics in New Zealand*,
https://doi.org/10.1007/978-3-030-18460-5

H.S. Dungey, C.B. Low, N.J. Ledgard & G.T. Stovold 2011a: Alternatives to *Pinus radiata* in the New Zealand high country: early growth and survival of *P. radiata, P. attenuata* and their F1 hybrid. NZJFS 41:61-69

62. C.J.A. Shelbourne, J.A. Zabkiewicz and P.A. Allan 1973a: Monoterpene composition in provenances of Pinus *muricata* planted in New Zealand.

63. C.J.A. Shelbourne 1973: Recent investigations of wood properties and growth performance in 'blue strain' *Pinus muricata*. NZ J. For. 19(1):13-45 1974a

40/6939 C.J.A. Shelbourne 1974b: *Pinus muricata* provenance variation in height and state of flushing of the terminal bud, 18 months from sowing

147/7052 C.J.A. Shelbourne 1976a? Two-year heights of 35 wind-pollinated progenies of *P. muricata*

180/7091 M.D. Wilcox, C.B. Low and M.H. Bannister 1978a: Assessment at age 6 years of the Pinus muricata gene pool experiment at Rotoehu forest.

R1312. C.J.A. Shelbourne, M.H. Bannister and M.D. Wilcox 1982a: Early results of provenance studies on *Pinus muricata* in New Zealand. NZ J. For. 27(1): 50-66.

24. I.J. Thulin, J.T. Miller 1966a: Provenance trials of *Pinus contorta* planted from 1958-1961; Establishment report.

27/6926 J.W. Hignett 1971a: Provenance variation in height and wood density in *Pinus contorta*

118/7025 J.T. Miller 1976a: Assessment of Pinus contorta provenance trials at Tara Hills.

38. J.T. Miller 1968a: The genetic improvement of *Pinus contorta* in New Zealand.

55. J.T. Miller 1971a: Provenance variation in growth rate and other characters in 6-year-old *Pinus contorta*.

R. 993. C.J.A. Shelbourne and J.T. Miller 1976b: Provenance variation in *Pinus contorta*. 6-year results from IUFRO seedlots in New Zealand. IUFRO XVI IUFRO World Congress Norway 1976. Proc. Div. II:140-145.

192/7118 O. Mohrdiek and J.T. Miller 1980a: Five-year assessment of open-pollinated progeny tests in *Pinus contorta*.

9. G.B. Sweet and M.P. Bollman 1962a: Provenance trial in Pinus ponderosa: A combined establishment and first assessment report.

40. I.J. Thulin and M.D. Wilcox 1970a: Ponderosa pine provenance trials (1960-61). Establishment report

47. M.D. Wilcox 1970a: Ponderosa pine provenance trials (1960-61) assessment six years after planting (1967).

R.D. Burdon and C.B. Low 1991a: Performance of *Pinus ponderosa* and *Pinus jeffreyi* provenances in New Zealand. Can. J. For. Res. 21:1401-1414.

R. 2115. J.T. Miller and C.J.A. Shelbourne 1984a: Sitka spruce provenance trials in New Zealand

R. 2621. R.D. Burdon and J.T. Miller 1995a: Alternative species revisited: categorisation and issues for strategy and research. NZ Forestry Aug. 1995.

IR 83 PROPPS 718 J.T. Miller 1978a: Species provenance trials of Sierra Redwood: establishment and first assessment.

PDF 21243 in SIDNEY same number. W.J. Libby 1999a: Observations of Giant Sequoia on South Island New Zealand.

15. I.J. Thulin 1963b Provenance trials of *Pinus attenuata*. Establishment report

M.P. Bollman 1966b: The height growth of a provenance trial in *Pinus attenuata* in Kaingaroa Forest Cpt 905.

H.S. Dungey, C.B. Low, N.J. Ledgard & G.T. Stovold 2011b: Alternatives to *Pinus radiata* in the New Zealand high country: early growth and survival of *P. radiata, P. attenuata* and their F1 hybrid. NZJFS 41:61-69

62. C.J.A. Shelbourne, J.A. Zabkiewicz and P.A. Allan 1973: Monoterpene composition in provenances of Pinus *muricata* planted in New Zealand.

63. C.J.A. Shelbourne 1973: Recent investigations of wood properties and growth performance in 'blue strain' *Pinus muricata*. NZ J. For. 19(1):13-45 1974c

40/6939 C.J.A. Shelbourne 1974d: *Pinus muricata* provenance variation in height and state of flushing of the terminal bud, 18 months from sowing

147/7052 C.J.A. Shelbourne 1976c? Two year heights of 35 wind-pollinated progenies of *P. muricata*
180/7091 M.D. Wilcox, C.B. Low and M.H. Bannister 1978b: Assessment at age 6 years of the Pinus muricata gene pool experiment at Rotoehu forest.
R1312. C.J.A. Shelbourne, M.H. Bannister and M.D. Wilcox 1982b: Early results of provenance studies on *Pinus muricata* in New Zealand. NZ J. For. 27(1): 50-66.
24. I.J. Thulin, J.T. Miller 1966b: Provenance trials of *Pinus contorta* planted from 1958-1961; Establishment report.
27/6926 J.W. Hignett 1971b: Provenance variation in height and wood density in *Pinus contorta*
118/7025 J.T. Miller 1976b: Assessment of Pinus contorta provenance trials at Tara Hills.
38. J.T. Miller 1968b: The genetic improvement of *Pinus contorta* in New Zealand.
55. J.T. Miller 1971b: Provenance variation in growth rate and other characters in 6-year-old *Pinus contorta*.
R. 993. C.J.A. Shelbourne and J.T. Miller 1976: Provenance variation in *Pinus contorta*. 6-year results from IUFRO seedlots in New Zealand. IUFRO XVI IUFRO World Congress Norway 1976. Proc. Div. II:140-145.
192/7118 O. Mohrdiek and J.T. Miller 1980b: Five year assessment of open-pollinated progeny tests in *Pinus contorta*.
9. G.B. Sweet and M.P. Bollman 1962b: Provenance trial in Pinus ponderosa: A combined establishment and first assessment report.
40. I.J. Thulin and M.D. Wilcox 1970b: Ponderosa pine provenance trials (1960-61). Establishment report
47. M.D. Wilcox 1970b: Ponderosa pine provenance trials (1960-61) assessment six years after planting (1967).
R.D. Burdon and C.B. Low 1991b: Performance of *Pinus ponderosa* and *Pinus jeffreyi* provenances in New Zealand. Can. J. For. Res. 21:1401-1414.
R. 2115. J.T. Miller and C.J.A. Shelbourne 1984b: Sitka spruce provenance trials in New Zealand
R. 2621. R.D. Burdon and J.T. Miller 1995b: Alternative species revisited: categorisation and issues for strategy and research. NZ Forestry Aug. 1995.
IR 83 PROPPS 718 J.T. Miller 1978b: Species provenance trials of Sierra Redwood: establishment and first assessment.
PDF 21243 in SIDNEY same number. W.J. Libby 1999b: Observations of Giant Sequoia on South Island New Zealand.

## Douglas-Fir Provenance and Breeding

14. G.B. Sweet 1963: Some provenance differences in *Pseudotsuga menziesii*. 1. Seed characteristics.
21. G.B. Sweet 1964a: The establishment of provenance trials in Douglas fir (*Pseudotsuga menziesii*) in New Zealand.
23. G.B. Sweet 1964b: The assessment six years after planting of a provenance trial in Douglas fir (1957 series).
G.B. Sweet 1965: Provenance differences in Pacific coast Douglas fir. Silvae Genetica 14(2): 46-56.
37. M.D. Wilcox 1968: The genetic improvement of Douglas fir in New Zealand
56. G.B. Sweet and M.P. Bollman 1971: Variation in seed yields per cone in Douglas fir in New Zealand.
59. C.J.A. Shelbourne, J.M. Harris, J.R. Tustin and I.D. Whiteside 1973: The relationship of timber stiffness to branch and stem morphology and wood properties in plantation-grown Douglas-fir in New Zealand.
55/6960 M.D. Wilcox 1974a: Stress grading study of Douglas-fir; effects of branch diameter and wood density on timber stiffness.

69. M.D. Wilcox 1974b: Douglas fir provenance variation and selection in New Zealand.
M.J.F. Lausberg, D.J. Cown, D.L. McConchie and J.H. Skipwith 1995a: Variation in some wood properties of *Pseudotsuga menziesii* provenances grown in New Zealand. NZJFS (25):
H. McConnon, R.L. Knowles and L.W. Hansen 2004: Provenance affects bark thickness in Douglas fir. NZJFS 34(1): 77-86
103/7010 T.G. Vincent and M.D. Wilcox 1976: Height and health of seven-year-old Mexican, Californian and Kaingaroa origin Douglas-fir at Kaingaroa and Gwavas forests.
Restarting the Douglas-fir breeding programme in 1988. (See Aus.For. later: Shelbourne, Low, Gea and Knowles 2005).
C.B. Low and L.D. Gea 1997: Estimation of genetic parameters for growth and form traits in Douglas-fir progeny test on four sites aged 23 years.
C.J.A. Shelbourne, C.B. Low, L.D. Gea, and R.L. Knowles 2007: Achievements in forest tree genetic improvement in Australia and New Zealand: 5: Genetic improvement of Douglas-fir in New Zealand. Aus. For. 70, 1 pp. 28-32
C.B. Low, N.J. Ledgard and C.J.A. Shelbourne 2012a: Early growth and form of coastal provenances and progenies of Douglas-fir at three sites in New Zealand. NZJFS 42:143-160
C.B. Low, C.J.A. Shelbourne and D.G. Henley 2012b: Effect of seed source of Douglas-fir at high-elevation New Zealand sites: performance at age eight years. NZJFS 42: 161-1
N.J. Ledgard and M.C. Belton 1985: Exotic trees in the Canterbury high country. NZJFS 15:
J.M. Harris 1966: The physical properties of NZ grown *Pinus* spp. other than *P. radiata* Symposium No. 7 1966

# Eucalypts

## Species and Provenance Research and Breeding Programs

10. G.B. Sweet 1962e: The future supply of high quality eucalypt seed for New Zealand. An analysis of the problems involved.
R. 40. I.J. Thulin and T. Faulds 1962?: Grafting of eucalypts. NZ J. For. 8(4):664-667.
R. 1379. M.D. Wilcox 1980: Genetic improvement of eucalypts in New Zealand. NZJFS 10(2): 3 -359.
R. 1295. M.D. Wilcox 1979: The ash group of eucalypts. NZJFS 9(2): 133-144.
R. 1298. M.D. Wilcox and I.J. Thulin 1979: Growth of *Eucalyptus regnans* in a plot at Rotorua. NZJF 9 (2): 166-169.
R. 1307. M.D. Wilcox, T. Faulds, T.G. Vincent and B.R. Poole *Eucalyptus regnans* F.Muell. Aust. For. Res. 10:169-84.
R. 1308. D.A. Rook, M.D. Wilcox, D.G. Holden and I.J. Warrington 1980: Provenance variation in frost tolerance of *Eucalyptus regnans* F. Muell. Aust. For. Res. 10:213-238.
M.I. Menzies, D.G. Holden, D.A. Rook and A.K. Hardacre 1981: Seasonal frost-tolerance of *Eucalyptus saligna, E. regnans* and *E. fastigata.* NZJFS (11):
M. Dick:1982 Leaf-inhabiting fungi of eucalypts in New Zealand. NZJFS (12):
R. 1503. M.D. Wilcox 1982a: Anthocyanin polymorphism in seedlings of *Eucalyptus fastigata* Deane et Maid. Aust. For. Res. 30:501-9.
R. 1604. M.D. Wilcox, T. Faulds and T.G. Vincent 1980a: Genetic improvement of *Eucalyptus saligna* Sm in New Zealand. IUFRO Symposium and Workshop, Aguas de Sao Pedro, Sao Paulo, Brazil Aug. 1980 (also published 1982 by FAO Rome in Forest Genetic Resources Information No 11).
M.D. Wilcox 1982b: Genetic variation in frost tolerance, early height growth rate and incidence of forking among and within provenances of *E. fastigata*. NZJFS ( 12):

R. 1605. M.D. Wilcox, D.A. Rook and D.G. Holden 1980b: Provenance variation in frost resistance of *Eucalyptus fastigata* Deane & Maid. IUFRO Symposium and Workshop, Aguas de Sao Pedro, Sao Paulo, Brazil. Aug. 1980. (also published 1982 by FAO Rome in Forest Genetic Resources Information No 11).

R. 1590. M.D. Wilcox 1982c: Preliminary selection of suitable provenances of *Eucalyptus regnans* for New Zealand. NZJFS 12(3): 468-479.

R. 1591. M.D. Wilcox 1982d: Selection of genetically-superior *Eucalyptus regnans* using family tests. NZJFS 12(3): 480-493.

FRI Bulletin No. 95. M.D. Wilcox, J.T. Miller, I.M. Williams and D.W. Guild 1985: *Eucalyptus* species trials in Longwood forest, Southland. Bulletin No. 95, FRI, NZFS, PB. Rotorua, NZ.

J.N. King and M.D. Wilcox 1988: Family tests as a basis for the genetic improvement of *Eucalyptus nitens* in New Zealand. NZJFS 18:

R.L. Hathaway and M. King 1986: Selection of *Eucalyptus* species for soil conservation planting in seasonally dry hill country. NZJFS 16:

R. 2053. M.D. Wilcox and R.L. Hathaway 1988: Use of Australian trees in New Zealand. Proc. AFDI International Forestry Conference for the Australian Bicentenary 1988, Albury, New South Wales, Australia Volume III.

R. 2467. P.G. Cannon and C.J.A. Shelbourne 1991: The New Zealand eucalypt breeding programme. IUFRO Symposium on Intensive Forestry: The Role of Eucalypts P2.0201 Productivity of eucalypts. Durban, South Africa 2-6 September, 1991

R. 2483. P.G. Cannon and C.J.A. Shelbourne 1993: Forward selection plots in breeding programmes with insect-pollinated tree species. NZJFS 23(1): 3-9.

J.N. King, R.D. Burdon and M.D. Wilcox 1993a: Provenance variation in New Zealand-grown *Eucalyptus delegatensis*. 1. Growth rates and form. NZJFS 23:

R. 2528. J.N. King, R.D. Burdon and G.D. Young 1993b: Provenance variation in New Zealand-grown *Eucalyptus delegatensis*. 2: Internal checking and other wood properties. NZJFS 23(3): 314-323.

M.J.F. Lausberg, K.F. Gilchrist and J.H. Skipwith 1995b: Wood properties of *Eucalyptus nitens* grown in New Zealand. NZJFS (25):

L.D. Gea, R.M. McConnochie and N.M.G. Borralho 1997a: Genetic parameters for growth and wood density traits in *Eucalyptus nitens* in New Zealand. NZJFS (27):

R. 2659. R.P. Kibblewhite, M.J.C. Riddell and C.J.A. Shelbourne 1998: Kraft fibre and pulp qualities of 29 trees of New Zealand grown *Eucalyptus nitens*. Appita 51(2): 114-121.

C.B. Low and C.J.A. Shelbourne 1999: Performance of *Eucalyptus globulus, E. maidenii, E. nitens*, and other eucalypts in Northland and Hawke's Bay at ages 7 and 11 years. NZJFS 29 (2): 274-288

C.J.A. Shelbourne, S.O. Hong, R.M. McConnochie and B. Pierce 1999a: Early results from trials of interspecific hybrids of *Eucalyptus grandis* with *E. nitens* in New Zealand. NZJFS 29(2): 251-262.

R. 2709. R.P. Kibblewhite and C.J. McKenzie 1999: Kraft fibre variation among 29 trees of 15-year-old *Eucalyptus fastigata* and comparison with *E. nitens*. Appita J. 52(3): 218-225

R.P. Kibblewhite, M.J.C. Riddell and C.J.A. Shelbourne 2000a: Variation in wood, kraft fibre, and handsheet properties among 29 trees of *Eucalyptus regnans*, and comparison with *E. nitens* and *E. fastigata*. NZJFS 30

R.P. Kibblewhite, B.L. Johnson and C.J.A. Shelbourne 2000b: Kraft pulp qualities of *Eucalyptus nitens, E. globulus* and *E. maidenii* at ages 8 and 11 years. NZJFS 30 (3): 347-357

C.J.A. Shelbourne, S.O. Hong, R.M. McConnochie and B. Pierce 1999b: Early results from trials of interspecific hybrids of *Eucalyptus grandis* with *E. nitens* in New Zealand. NZJFS 29(2): 251-262.

R.B. McKinley, C.J.A. Shelbourne, J.M. Harris and G.D. Young 2000: Variation in whole-tree basic wood density for a range of plantation species grown in New Zealand. NZJFS 30 (3): 436-446.

C.J.A. Shelbourne, B.T. Bulloch, R.L. Cameron and C.B. Low 2000a: Results of provenance testing *Acacia dealbata, A. mearnsii* and other Acacias at ages 7 and 5 years in New Zealand. NZJFS 30 (3): 401-421.

C.J.A. Shelbourne, C.B. Low and P.J. Smale 2000b: Eucalypts for Northland: 7- to 11-year results from trials of nine species at four sites. NZJFS 30 (3): 366-383.

R. 2813. R.B. McKinley, C.J.A. Shelbourne, C.B. Low, B. Penellum and M.O. Kimberley 2002: Wood properties of young *Eucalyptus nitens, E. globulus* and *E. maidenii* in Northland, New Zealand. NZJFS 32(3):334-356.

C.J.A. Shelbourne, I.D. Nicholas, R.B. McKinley, C.B. Low, R.M. McConnochie and M.J.F. Lausberg 2002a: Wood density and internal checking of young *Eucalyptus nitens* in New Zealand as affected by site and height up the tree. NZJFS 32

C.J.A. Shelbourne, I.D. Nicholas, R.B. McKinley, C.B. Low, R.M. McConnochie and M.J.F. Lausberg 2002b: Wood density and internal checking of young *Eucalyptus nitens* in New Zealand as affected by site and height up the tree. NZJFS 32

C.J.A. Shelbourne, B.T. Bulloch, C.B. Low and R.M. McConnochie 2002c: Performance to age 22 years of 49 eucalypt species in the Wairarapa district, New Zealand, and results from other **trials.** NZJFS 32(2): 256-278

H.M. McKenzie, J.C.P. Turner and C.J.A. Shelbourne 2003a: Processing young plantation-grown *Eucalyptus nitens* for solid-wood products. 1. Individual-tree variation in quality and recovery of appearance-grade lumber and veneer. NZJFS 33(1): 62-78

H.M. McKenzie, C.J.A. Shelbourne, M.O. Kimberley, R.B. McKinley and R.A.J. Britton 2003b: Processing young plantation-grown *Eucalyptus nitens* for solid-wood products. 2. Predicting product quality from tree, increment core, disc and 1 m. billet properties. NZJFS 33(1): 79-113

H.M. McKenzie, C.J.A. Shelbourne, M.O. Kimberley, R.B. McKinley and R.A.J. Britton 2003c: Processing young plantation-grown *Eucalyptus nitens* for solid-wood products. 2. Predicting product quality from tree, increment core, disc and 1 m. billet properties. NZJFS 33(1): 79-113

L.D. Gea, R.M. McConnochie and S. Wynyard 2007: Parental reconstruction for breeding, deployment and seed orchard management of *Eucalyptus nitens*. NZJFS 37 ( 1): 23-36

T.G. Jones, R.M. McConochie, C.J.A. Shelbourne and C.B. Low 2010: Sawing and grade recovery of 25-year-old *Eucalyptus fastigata, E. globoidea, E. muelleriana* and *E. pilularis*. NZJFS 40: 19-31

## *Cupressus* and Other Conifers

20. A.J. Carruthers 1964: Establishment plan for a provenance trial of *Abies grandis*.

L.E. Fung 1993: Wood properties of New Zealand-grown *Cunninghamia lanceolata*. NZJFS 23:

Bannister, M.H. 2009: Variation in seedlings of *Cupressus lusitanica*. NZJFS 39: 57-64

C.B. Low, H.M. McKenzie, C.J.A. Shelbourne and L.D. Gea 2005a: Sawn-timber and wood properties of 21-year-old *Cupressus lusitanica, C. macrocarpa* and *Chamaecyparis nootkatensis x C. macrocarpa* hybrids. Part 1. Sawn timber performance. NZJFS 35

C.B. Low, H.M. McKenzie, C.J.A. Shelbourne and L.D. Gea 2005b: Sawn-timber and wood properties of 21-year-old *Cupressus lusitanica, C. macrocarpa* and *Chamaecyparis nootkatensis x C. macrocarpa* hybrids. Part 1. Sawn timber performance. NZJFS 35

39587 C.B. Low and S.J. Gatenby 2006: Assessment results of the 1985 cypress species trial at 3 locations.

L.D. Gea and C.B. Low 1997: Genetic parameters for growth, form and canker resistance of *Cupressus macrocarpa* in New Zealand. NZJFS (27):

R. 374. I.J. Thulin 1969a: Breeding *Pinus radiata* through seed improvement and clonal afforestation. Second World Consultation on Forest Tree Breeding, Washington, 7-12 Aug. 1969.

## Clonal Forestry

R. 374. I.J. Thulin 1969b: Breeding *Pinus radiata* through seed improvement and clonal afforestation. Second World Consultation on Forest Tree Breeding, Washington, 7-12 Aug. 1969.

C.J.A. Shelbourne 1969a: Tree breeding methods. Technical Paper 55, Forest Research Institute, New Zealand Forest Service, Wellington.

R. 766. C.J.A. Shelbourne and I.J. Thulin 1974: Early results from a clonal selection and testing programme with radiata pine. N.Z. J. For. Sci. 4 (2); 387-398.

I.J. Thulin, A. Firth, T.G. Vincent and C.J.A. Shelbourne 1973-1977: Cutting propagation of clones selected in 1968-planted full-sib block plantings at Cpt. 1350, Kaingaroa forest with subsequent establishment of hedges.

R.D. Burdon and J.M. Harris 1972: Wood density of radiata pine clones on four different sites. NZJFS 3 (1972).

R. 768. R.D. Burdon and C.J.A. Shelbourne 1974: The use of vegetative propagules for obtaining genetic information. NZJF 4 (2): 418-425.

M.J. Carson 1986: Advantages of clonal forestry for *Pinus radiata*. Real or imagined? NZJFS 6 (1986).

R.D. Burdon 1988: When is cloning on an operational scale appropriate? Proc. IUFRO Conf Breeding tropical trees. Pattaya, Thailand 28 Nov. - 3 Dec. 1988

R. 2402. C.J.A. Shelbourne 1992a: Genetic gains from different kinds of breeding population and seed or plant production population. South African Forestry Journal 160 March 1992.

R.D. Burdon, R.D. Gaskin, C.B. Low and G. Zabkiewicz 1992: Clonal repeatability of monoterpene composition of cortical oleoresin of *Pinus radiata*. NZJFS 22 (1992).

R. Beauregard, R. Gazo, M.O. Kimberley, J. Turner, S. Mitchell and C.J.A. Shelbourne 1999: Clonal variation in the quality of radiata pine random width boards. Wood and Fibre Science, 31(3) 1999.

C.J.A. Shelbourne 1997: Genetics of adding value to the end-products of radiata pine. Proc. IUFRO Genetics of Radiata Pine, Rotorua, New Zealand, 1-4 December 1997. (FRI Bulletin No. 203)

B.S. Baltunis, H. Wu, H.S. Dungey, T.J. Mullin and J.T. Brawner 2009: Comparisons of genetic parameter and clonal value predictions from clonal trials and seedling base population trials of radiata pine. Tree Genetics and Genomes. Jan. 5: 269-278.

M.J. Carson, S.D. Carson and C. Te Rini 2005: Successful varietal forestry with radiata pine in New Zealand. NZ J. For. May 2005 60 -1.

B.C. Baltunis and J.T. Brawner 2010: Clonal stability in *Pinus radiata* across New Zealand and Australia. 1. Growth and form traits. New Forests DOI 10 1007

## Breeding Theory

C.J.A. Shelbourne 1969b: Tree breeding methods. Technical Paper 55, Forest Research Institute, New Zealand Forest Service, Wellington.

41. C.J.A. Shelbourne and F.R.M. Cockrem 1969a: Progeny and clonal test designs for New Zealand's tree breeding programme.

R. 772. C.J.A. Shelbourne 1973b: Problems and prospects in the improvement of forest tree species. Proc. 2$^{Nd}$ General Congress, SABRAO, New Delhi 1973. reprinted from Indian J. Genet., 34A 1974.

R. 531. R.D. Burdon and C.J.A. Shelbourne 1971a: Breeding populations for recurrent selection: conflicts and possible solutions. NZJFS 1(2):1174-193.

R. 548 C.J.A. Shelbourne 1971b: Planning breeding programs for tropical conifers grown as exotics. IUFRO (section 22) Gainesville, Fla. 1971. Symposium on "Selection and breeding to improve some tropical conifers"

R. 683 C.J.A. Shelbourne 1972a: Genotype-environment interaction: its study and its implications in forest tree improvement. IUFRO Genetics-Sabrao Joint Symposia.Tokyo, October 1972.

R. 1021. R.D. Burdon, C.J.A. Shelbourne and M.D. Wilcox 1977a: Advanced selection strategies. Third World Consultation on Forest Tree Breeding, Canberra and Rotorua.

R. 1089. R.D. Burdon 1977a: Genetic correlation as a concept for studying genotype-environment interaction in forest tree breeding. Silvae Genetica 26(5-6):145-228.

R. 1296. R.D. Burdon 1981a: Generalisation of multi-trait selection indices using information from several sites. NZJFS 9(2): 145-152.

R. 1518. R.D. Burdon 1982a: Selection indices using information from multiple sources for the single trait case. Silvae Genetica 31(2-3): 81-85.

R. 1599. R.D. Burdon and G. Namkoong 1983a: Short note: Multiple populations and sublines. Silvae Genetica 32(5-6):221-222.

R. 1640. R.D. Burdon 1982b: Breeding for productivity- jackpot or will-o-the-wisp? Proc. North American Forest Biology Workshop, University of Kentucky, Lexington, Ke., July 1982.

R. 1736. G.B. Sweet, and R.D. Burdon 1983a: The radiata pine monoculture: an examination of the ideologies. NZ J. For. 28(3): 325-26.

D.V. Shaw and J.W. Hood 1985a: Maximising gain per effort by using clonal replication in genetic tests. TAG 71:392-399.

R.D. Burdon 1986a: Breeding long-lived perennials-frustration, temptations, opportunities. Plant Breeding Symposium DSIR 1986 Agron. Soc. NZ special pub. no. 5.

R.D. Burdon 1989a: Early selection in tree breeding: principles for applying index selection and inferring input parameters. Can. J. For. Res. 19: 499-504.

C.J.A. Shelbourne 1992b: Genetic gains from different kinds of breeding populations and seed or plant production populations. S. Afr. For. J. 160:49-6

R. 2453. M.J. Carson, T.G. Vincent and A. Firth 1992a: Control-pollinated and meadow seed orchards of radiata pine. Mass Production Technology for Genetically Improved Fast growing Forest Tree Species. Bordeaux 14-18 Sept. 1992. AFOCEL Paris 1992

R. 2503 R.D. Burdon 1992a: Testing and selection: strategies and tactics for the future. pp. 249-60 Proc. IUFRO Conf. Oct. 9-18 1992, Catagena and Cali, Colombia. Sponsored by CAMCORE.

T.L. White, G.R. Hodge and G.L. Powell 1993a: An advanced genetic improvement plan for slash pine in the southeastern United States. Silvae Genetica 42 (6): 359-370.

R. 2992 R.D. Burdon 1995a: New directions in tree breeding: some questions of what we should seek, and how to manage the genetic resource. CTIA/WFGA meeting, Victoria, BC August 1995.

R. 2659. L.D. Gea, D. Lindgren, C.J.A. Shelbourne and T. Mullin 1997b: Complementing inbreeding coefficient information with status number: implications for structuring breeding populations. NZJFS 27(3): 255-271.

C.J.A. Shelbourne, L.A. Apiolaza, K.J.S. Jayawickrama and C.T. Sorensson 1997a: Developing breeding objectives for radiata pine in New Zealand. Proc. IUFRO '97 Genetics of Radiata Pine, Rotorua, NZ, Dec. 1 - 4 1997

O. Rosvall, D. Lindgren and T.J. Mullin 1998a: Sustainability, robustness and efficiency of a multi-generation breeding strategy based on within-family clonal selection. Silvae Genetica 47 (5/6): 307-320.

R. 2745. R.D. Burdon 2001a: Genetic aspects of risk: species diversification, genetic. management and genetic engineering. NZ J. Forestry, Feb. 2001:20-25.

R. 2842. R.D. Burdon and S. Kumar 2004a: Forwards versus backwards selection: trade-offs between expected genetic gain and risk avoidance. NZJFS 34(1): 3-21.

R. 2847. R.D. Burdon, R.P. Kibblewhite, J.F. Walker, R. Evans and D.J. Cown 2004a: Juvenile versus mature wood: a new concept, orthogonal to corewood versus outerwood, with special reference to *Pinus radiata* and *P. taeda*. Forest Science 50(4): 399-415

R. 2745. R.D. Burdon 2001b: Genetic aspects of risk: species diversification, genetic. management and genetic engineering. NZ J. Forestry, Feb. 2001:20-25.

## Radiata Breeding

C.J.A. Shelbourne 1969c: Tree breeding methods. Technical Paper 55, Forest Research Institute, New Zealand Forest Service, Wellington.

41. C.J.A. Shelbourne and F.R.M. Cockrem 1969b: Progeny and clonal test designs for New Zealand's tree breeding programme.

R. 772. C.J.A. Shelbourne 1973c: Problems and prospects in the improvement of forest tree species. Proc. 2[Nd] General Congress, SABRAO, New Delhi 1973. reprinted from Indian J. Genet., 34A 1974.

R. 531. R.D. Burdon and C.J.A. Shelbourne 1971b: Breeding populations for recurrent selection: conflicts and possible solutions. NZJFS 1(2):1174-193.

R. 548 C.J.A. Shelbourne 1971c: Planning breeding programs for tropical conifers grown as exotics. IUFRO (section 22) Gainesville, Fla. 1971. Symposium on "Selection and breeding to improve some tropical conifers"

R. 683 C.J.A. Shelbourne 1972b: Genotype-environment interaction: its study and its implications in forest tree improvement. IUFRO Genetics-Sabrao Joint Symposia.Tokyo, October 1972.

R. 1021. R.D. Burdon, C.J.A. Shelbourne and M.D. Wilcox 1977b: Advanced selection strategies. Third World Consultation on Forest Tree Breeding, Canberra and Rotorua.

R. 1089. R.D. Burdon 1977b: Genetic correlation as a concept for studying genotype-environment interaction in forest tree breeding. Silvae Genetica 26(5-6):145-228.

R. 1296. R.D. Burdon 1981b: Generalisation of multi-trait selection indices using information from several sites. NZJFS 9(2): 145-152.

R. 1518. R.D. Burdon 1982c: Selection indices using information from multiple sources for the single trait case. Silvae Genetica 31(2-3): 81-85.

R. 1599. R.D. Burdon and G. Namkoong 1983b: Short note: Multiple populations and sublines. Silvae Genetica 32(5-6):221-222.

R. 1640. R.D. Burdon 1982d: Breeding for productivity- jackpot or will-o-the-wisp? Proc. North American Forest Biology Workshop, University of Kentucky, Lexington, Ke., July 1982.

R. 1736. G.B. Sweet, and R.D. Burdon 1983b: The radiata pine monoculture: an examination of the ideologies. NZ J. For. 28(3): 325-26.

D.V. Shaw and J.W. Hood 1985b: Maximising gain per effort by using clonal replication in genetic tests. TAG 71:392-399.

R.D. Burdon 1986b: Breeding long-lived perennials-frustration, temptations, opportunities. Plant Breeding Symposium DSIR 1986 Agron. Soc. NZ special pub. no. 5.

R.D. Burdon 1989b: Early selection in tree breeding: principles for applying index selection and inferring input parameters. Can. J. For. Res. 19: 499-504.

C.J.A. Shelbourne 1992c: Genetic gains from different kinds of breeding populations and seed or plant production populations. S. Afr. For. J. 160:49-6

R. 2453. M.J. Carson, T.G. Vincent and A. Firth 1992b: Control-pollinated and meadow seed orchards of radiata pine. Mass Production Technology for Genetically Improved Fast growing Forest Tree Species. Bordeaux 14-18 Sept. 1992. AFOCEL Paris 1992

R. 2503 R.D. Burdon 1992b: Testing and selection: strategies and tactics for the future. pp. 249-60 Proc. IUFRO Conf. Oct. 9-18 1992, Catagena and Cali, Colombia. Sponsored by CAMCORE.

T.L. White, G.R. Hodge and G.L. Powell 1993b: An advanced genetic improvement plan for slash pine in the southeastern United States. Silvae Genetica 42 (6): 359-370.

R. 2992 R.D. Burdon 1995b: New directions in tree breeding: some questions of what we should seek, and how to manage the genetic resource. CTIA/WFGA meeeting, Victoria, BC August 1995.

R. 2659. L.D. Gea, D. Lindgren, C.J.A. Shelbourne and T. Mullin 1997c: Complementing inbreeding coefficient information with status number: implications for structuring breeding populations. NZJFS 27(3): 255-271.

C.J.A. Shelbourne, L.A. Apiolaza, K.J.S. Jayawickrama and C.T. Sorensson 1997b: Developing breeding objectives for radiata pine in New Zealand. Proc. IUFRO '97 Genetics of Radiata Pine, Rotorua, NZ, Dec. 1 - 4 1997

O. Rosvall, D. Lindgren and T.J. Mullin 1998b: Sustainability, robustness and efficiency of a multi-generation breeding strategy based on within-family clonal selection. Silvae Genetica 47 (5/6): 307-320.

R. 2745. R.D. Burdon 2001c: Genetic aspects of risk: species diversification, genetic. management and genetic engineering. NZ J. Forestry, Feb. 2001:20-25.

R. 2842. R.D. Burdon and S. Kumar 2004b: Forwards versus backwards selection: trade-offs between expected genetic gain and risk avoidance. NZJFS 34(1): 3-21.

R. 2847. R.D. Burdon, R.P. Kibblewhite, J.F. Walker, R. Evans and D.J. Cown 2004b: Juvenile versus mature wood: a new concept, orthogonal to corewood versus outerwood, with special reference to *Pinus radiata* and *P. taeda*. Forest Science 50(4): 399-415

R. 2745. R.D. Burdon 2001d: Genetic aspects of risk: species diversification, genetic. management and genetic engineering. NZ J. Forestry, Feb. 2001:20-25.

# Index





C. J. A. Shelbourne, M. Carson, *Tree Breeding and Genetics in New Zealand*,
https://doi.org/10.1007/978-3-030-18460-5

The manufacturer's authorised representative in the EU is Springer Nature Customer Service Centre GmbH, Europaplatz 3, 69115 Heidelberg, Germany. If you have any concerns regarding our products, please contact ProductSafety@springernature.com

Printed and bound by CPI Group (UK) Ltd, Croydon, CR0 4YY
30/09/2026
02225693-0003